DESIGN DATA FOR REINFORCED PLASTICS

This book is dedicated to
Margaret, Mark and Duncan
and
Pauline, David and Helen

DESIGN DATA FOR REINFORCED PLASTICS

A guide for engineers and designers

Neil L. Hancox

Project Manager
AEA Technology
Harwell, UK

and

Rayner M. Mayer

Consultant
Sciotech
Yateley, UK

CHAPMAN & HALL
London · Glasgow · New York · Tokyo · Melbourne · Madras

Published by Chapman & Hall, 2–6 Boundary Row, London SE1 8HN

Chapman & Hall, 2–6 Boundary Row, London SE1 8HN, UK

Blackie Academic & Professional, Wester Cleddens Road, Bishopbriggs, Glasgow G64 2NZ, UK

Chapman & Hall Inc., One Penn Plaza, 41st Floor, New York NY 10119, USA

Chapman & Hall Japan, Thomson Publishing Japan, Hirakawacho Nemoto Building, 6F, 1-7-11 Hirakawa-cho, Chiyoda-ku, Tokyo 102, Japan

Chapman & Hall Australia, Thomas Nelson Australia, 102 Dodds Street, South Melbourne, Victoria 3205, Australia

Chapman & Hall India, R. Seshadri, 32 Second Main Road, CIT East, Madras 600 035, India

---

First edition 1994

Typeset in Sabon 10.5/12.5 by Best-set Typesetter Ltd, Hong Kong
Printed in Great Britain by Clays Ltd, Bungay, Suffolk

ISBN 0 412 493209

A catalogue record for this book is available from the British Library

Library of Congress Cataloging-in-Publication data
Hancox, N.L.
Design data for reinforced plastics: a guide for engineers and designers/Neil Hancox and Rayner M. Mayer. – 1st ed.
p. cm.
Includes index.
ISBN 0-412-49320-9
1. Fiber reinforced plastics. I. Mayer, Rayner M. II. Title.
TA455.P5H34 1993
620.1′923—dc20 93-12190
CIP

∞ Printed on permanent acid-free text paper, manufactured in accordance with the proposed ANSI/NISO Z 39.48-199X and ANSI Z 39.48-1984

CONTENTS

PREFACE

Fibre-reinforced plastics are a group of materials whose properties can only be determined after a component or structure has been fabricated. This poses both a challenge and an opportunity for those who would use these materials:

- a challenge because one has to surmise or assume a certain set of properties in order to undertake the design;
- an opportunity because it is possible to place the reinforcement so that it balances the imposed loads exactly.

This book addresses the challenge, because the data required for design are scattered far and wide and are not otherwise available in a form in which they can be used directly.

The assembled data are believed to be characteristic of the principal materials used for general purpose mechanical engineering. This book is neither a compendium nor a database and so where gaps appear it is either because the data do not exist, or the authors have been unable to locate them.

If this book is found to be useful, then we hope that others may be encouraged to fill the gaps in our knowledge or submit data, which could be used in subsequent editions.

Whilst the book is self-contained, the accompanying volume on design principles (Mayer, R.M. (ed.) (1993) *Design with Reinforced Plastics*, Design Council, London) may need to be consulted if readers are not acquainted with the design process.

If this book helps to encourage others to design with these materials, then the authors' labours will have been well rewarded.

Rayner Mayer
Yateley
and
Neil Hancox
Abingdon,
June 1993.

LAYOUT OF THE BOOK

The first three chapters discuss the primary design considerations, the material selection process for composites and the properties of the basic constituents. These chapters should enable the designer to establish a broad perspective of what might be possible when such constituents are incorporated to form a composite.

Data on short-term mechanical properties of composites are presented in the next three chapters. A clear distinction is made between random, fabric and aligned reinforcement and the properties that these produce.

A further four chapters consider properties associated with the performance of a composite under various conditions, including the effects of impact, fire, environmental conditions and long-term loading, either continuous or cyclic.

Each of the major manufacturing processes is briefly described, together with a listing of the relevant data location. Quality control of processing is discussed in the final chapter.

HOW TO USE THIS BOOK

This book can be used in various ways:

For readers conversant with composites, key words or locators, placed in the margin, enable data to be located. The search may either be for a suitable material, or for a required property level.

If more basic information is required first, Chapter 2 should be consulted for the initial selection of constituents and Chapter 3 for specific properties of fibres and resins. Resin properties are amplified further in Chapter 4.

The classification of mechanical properties is by resin type in Chapter 4 and by fibre type in Chapter 5. This is consistent with the way in which the materials are used. The influence of processing on both material selection and properties is discussed in Chapter 11, and may restrict the choices set out in Chapters 4 and 5.

Chapters 6 to 10 describe other properties which need to be considered as part of the design process, whilst Chapter 12 outlines aspects of quality assurance to ensure compliance with the design.

If, as is often necessary, the design has to be iterated, then the material selection may need to be reconsidered, and the available data reviewed. Databases and manufacturers' information can be used to supplement what is set out here.

DISCLAIMER

The data here are provided in good faith, but neither we nor the original providers of the data are able to accept responsibility for the accuracy of any of the information included, or for any of the consequences that may arise from use of the data or designs or constructions based on any of the information supplied or materials described. The inclusion or omission of a particular material in no way implies anything about its performance with respect to other materials.

COPYRIGHT

ACKNOWLEDGEMENTS

This guide could not have been written without the contribution of many people throughout the industry and academia.

In particular, we would like to thank:

George Jeronimidis for providing encouragement and advice throughout the project;

David and Christine Wright of RAPRA Technology for advice and encouragement in undertaking this project and for a range of services provided in searching the literature through the RAPRA bibliographic base, which forms the starting point of this project;

Les Norwood (Scott Bader), Ragnar Arvesen (Jotun Polymer), Gerard Creux (Vetrotex), Jacques Gerard (OCF) and Murray Orpin (BP Chemicals) for in-depth discussions and the supply of data, much of which has not been previously published;

Marc Scudamore and Thomas Madgwick for advice and discussions on the fire performance of composites;

Andrew Walters for undertaking much of the background research;

The Materials Information Service of the Design Council;

Gill Money and Alan Thompson for assessing the manuscript and their helpful comments.

This guide could not have been researched and written without funding from the Materials Matter Program of the Department of Trade and Industry.

THE DATASETS

We have contacted, or attempted to contact, the author or lead author and publisher of every source from which we have quoted, or in the case of trade data the organization involved, to obtain permission to publish the data. Apart from the acknowledgements we make in the text to these people, we should also like to thank:

ASME for permission to publish from *Journal of Engineering Materials Technology*,

ASTM for permission to publish from *Journal of Composites Technology and Research* and various STPs, copyright ASTM,

Butterworth–Heinemann Ltd for permission to publish from *Composites*,

Elsevier Applied Science Ltd for permission to publish from *Composites Science and Technology*, *Composites Structures* and *ICCM* 6,

IEE Publishing Department and ERA Technology Ltd for permission to quote data for Kevlar composites in Table 7.8,

Institution of Mechanical Engineers for permission to publish from ME Conference, London 1985,

MMFG, Virginia, USA for permission to quote data in Table 7.3 for vinyl ester E glass fibre rod,

Owens Corning Fiberglas for permission to publish from *Design Data* by J. Quinn,

PRI (now IoM), London, for permission to reproduce Figures 5.9, 5.10 and 5.11,

Society for the Advancement of Materials and Process Engineering for permission to publish from *SAMPE Quaterly* and 33rd International SAMPE Symposium.

If any individual or organization should know of any copyright or source that has not been expressly acknowledged would they please write with details to the publisher at the London address given.

TRADEMARKS AND REGISTERED TRADE NAMES

The following registered trademarks are the properties of the organizations listed. Our use of them in this book in no way implies any transference of ownership from their current holders.

FIBRES

Glass fibres

E-CR GLAS and *S-2* glass are the registered trademarks of Owens Corning Fiberglas,
R glass is the registered trademark of Verotex International,
Equerove is the registered trademark of Pilkingtons.

Aramid fibres

Kevlar 129, H_p, *49* and H_m are the registered trademarks of du Pont,
Twaron is the registered trademark of Akzo,
Technora is the registered trademark of Teijin.

Oriented polyethylene fibres

Dyneema is the registered trademark of DSM,
Spectra 900, *1000* are the registered trademarks of Allied Signal.

Carbon fibres

HTA, *HM35*, *HM45*, *IM600* are the registered trademarks of Akzo Fibres,
P25, *P55S*, *P75S*, *P100S*, *P120S* are the registered trademarks of Amoco Performance,
GY70, *80*, *40–700* are the registered trademarks of BASF Structural Materials,
35-7, *50-8B* are the registered trademarks of BP Chemicals Hitcow,
HM400, *XA*, *IM* are the registered trademarks of Courtaulds,
55 is the registered trademark of Fiber Materials Inc.,
AS2, *AS4*, *IM7* are the registered trademarks of Hercules Advanced Materials,
KI39, *K137*, *T-A*, *MM-G*, *321* are the registered trademarks of Mitsubishi Kasei Corporation,

XN70 is the registered trademark of Nippon Petroleum,
IM400, *IM500* are the registered trademarks of Toho Rayon
FT700, *FT500* are the registered trademarks of Tonen Corporation,
T300, *T400H*, *T800H*, *T1000G*, *T1000*, *M35J*, *40J*, *46J*, *50J*, *55J*, *60J* are registered trademarks of Toray Industries.

Alumina fibres

Safimax SD, *LD* are the registered trademarks of ICI,
Saphikon is the registered trademark of Saphikon,
Altex is the registered trademark of Sumitomo.

Alumina/boria/silica fibres

Nextel 440, *480*, *312* are the registered trademarks of 3M.

Silicon carbide fibres

Sigma BP is the registered trademark of BP Metal Composites,
Nicalon is the registered trademark of Nippon Corporation,
Tyranno is the registered trademark of Ube Industries.

RESINS

Derakane is the registered trademark of Dow Chemicals,
Lexan and *Noryl* are the registered trademarks of General Electric,
Radel and *Torlon* are the registered trademarks of Amoco,
Victrex and *APC2* are the registered trademarks of ICI,
Crystic is the registered trademark of Scott Bader,
Epon and *Compimide* are the registered trademarks of Shell,
Fiberite is the registered trademark of Narmco.

If we have omitted a registered trademark from this list please inform the publisher at the London address given. Omission from this list in no way implies any loss of property on the part of the current owner.

GLOSSARY

Note: This glossary is self-consistent with that proposed by the ASM International Handbook Committee as set out in Volume 2 of the Engineered Materials Handbook, published by ASM, Metals Park, Ohio, USA

ABS	Acrylonitrile butadiene styrene resin.
Accelerator	A material that, when mixed with a catalyst or resin, speeds up the curing process.
Additive	A substance added to the resin to polymerize it such as accelerator, initiator, or catalyst; or to improve resin properties such as filler or flame retardant.
AFNOR	Association Français de Normalisation.
Aramid fibre	Organic-based high performance fibre.
Aspect ratio	Ratio of fibre length to diameter.
ASTM	American Society for Testing and Materials.
ATH	Aluminium trihydrate.
Autoclave moulding	A process in which both heat and pressure is applied to a composite placed in an autoclave.
Barcol hardness	Micro-hardness indentation technique.
B-staging	Partial reaction of a thermosetting resin.
Binder	A compound applied to fibre mat or preforms to bond the fibres before moulding or laminating.
BMC	Bulk moulding compound (see compounds).
BPF	British Plastics Federation.
BS	British Standards.
BSI	British Standards Institute.
BVID	Barely visible impact damage.
CAA	Civil Aviation Authority, UK.
C-glass fibres	A glass used for its chemical stability in corrosive environments.
Catalyst	A material that when added in small quantities increases the rate of cure of a resin.
CEC	Commission of the European Community.
CEN	Comite Européen de Normalisation.
Centrifugal moulding	A process in which chopped fibres, impregated with resin, are sprayed into the inside of a mould which is rotated to consolidate the mixture.
CFM	Continuous filament mat.

COM	Contact moulding fabrication process.
Compound	An intimate mixture of a resin with other ingredients such as catalysts, fillers, pigments and fibres – usually containing all ingredients necessary for the finished product.
Composite	A generic term to describe the mixture of a fibrous reinforcement and resin matrix.
Consolidation	A process by which the fibre/resin mixture is compressed by various processes such as a roller, vacuum or pressure to eliminate air bubbles and achieve desired properties.
Contact moulding	A process in which the composite is laid up on a mould either by hand or by spraying short length fibres impregated with resin.
CRM	Continuous random mat.
CFRP	Carbon fibre-reinforced plastic.
CSM	Chopped strand mat.
Cure	The process of cross-linking a plastic material to produce a rigid, solid object.
Curing agents	Materials added to resins to make them cure.
D-glass fibres	Glass with good dielectric properties.
Delamination	Separation between two or more layers of a composite due either to incorrect processing or subsequent degradation during use.
Denier	A numbering system for yarns and filaments in which the yarn number is equal to the weight in grams of 9000 metres. The lower the denier, the finer the yarn.
DIN	Deutsches Institut für Normung.
DMC	Dough moulding compound (see compound).
Drape	The ability of a fabric to conform to a shape or surface.
E-glass fibre	Glass fibre in general use.
E-CR glass fibre	A type of E-glass fibre with enhanced chemical resistance to corrosion.
EFTA	European Free Trade Association.
EN	European standard.
Epoxy resin	A thermosetting resin widely used with high performance fibres.
FAA	Federal Aviation Authority, USA.
Fibre	A material in filamentary form having a small diameter compared with its length.
Fibre content	The amount of fibre present in a composite usually expressed as a percentage volume or weight fraction.

Filament	See fibre.
Filament winding	A process which involves winding fibres or tapes onto a mandrel – these are generally pre-impregnated with resin.
Filler	An inert material added to a resin to alter its properties or to lower cost or density – generally in the form of a fine powder.
Finish	Materials used to coat filament bundles – usually contains a coupling agent to to improve the fibre to resin bond, a lubricant to prevent abrasion and a binder to preseve integrity of a filament bundle.
Finished items	Items made of fibre-reinforced plastic, which are fully cured – such items generally consist of standard sections such as rod, bar, tube, channel or plate and are usually available ex stock from suppliers.
Foam cores	A foamed resin created by using a foaming agent – rigid foams are useful as core materials in sandwich panels between stiffer outer layers.
FRP	Fibre-reinforced plastic.
G_1, G_2, G_3	Fracture energies.
Gel coat	The surface layer of a moulding used to improve surface appearance or properties – applied using a quick-setting resin.
Gel time	The time period from mixing of the curing agent with the resin until the mixture is sufficiently viscous that it does not readily flow.
Geometric efficiency	Measure of the directionality of the fibre reinforcement.
Glass transition temperature	Temperature at which significant changes occur in properties of cured resin due to enhanced molecular mobility.
GMT	Glass mat thermoplastic.
GRP	Glass-reinforced plastic.
Hardener	A substance added to a resin to promote curing.
HDT	Heat distortion temperature (Note: varies according to load applied during the test).
HM fibre	High modulus version of a fibre.
HS fibre	High strength version of a fibre.
Hybrid	A composite containing two (or more) types of reinforcement.
IEC	International Electro-technical Commission.
ILSS	Interlaminar shear stress.
IM	Injection moulding process.

IMC	Injection moulding compound (see compound).
Impregnation	The process of introducing resin into filament bundles or fabric laid up in a mould.
Initiator	A substance which provides a source of chemical agents to promote curing.
Injection	A process used for introducing a liquid resin (or heat softened thermoplastic) into a mould.
Injection moulding	A process which involves injecting a formulated moulding compound into a mould.
Interface	The area between the fibre and the matrix.
ISO	International Standards Organisation; International standard.
JIS	Japanese industrial standard.
Kevlar	A type of aramid fibre.
K	Stress intensity factor.
LEFM	Linear elastic fracture mechanics.
Mandrel	A form of mould (usually associated with a component having cylindrical symmetry).
Mat	A material consisting of randomly orientated fibre bundles which may be chopped or continuous, loosely held together with a binder or by needling.
Matrix	The resin in which the fibrous reinforcement is embedded.
Methacrylate	A thermosetting resin.
Mould	The tooling in which the composite is placed to give the correct shape to the article while the resin cures.
Moulding	An article manufactured in a mould.
Moulding compounds	A type of compound suitable for moulding (see compound).
Moulding process	The process by which an article is made in a mould.
NDE, NDT	Non-destructive evaluation, non-destructive testing.
Needled mat	A fabric which consists of short fibres felted together with a needle loom – a carrier may or may not be used.
NPG-polyester	A type of polyester resin formed using neopentyl glycol.
Oxygen index	A method for measuring flame spread.

PA	Polyamide resin (commonly known as nylon).
PAI	Polyamide imide resin.
PAS	Polyaryl sulphone resin.
PC	Polycarbonate resin.
PE	Polyethylene fibre or resin.
PEEK	Polyether ether ketone resin.
PEI	Polyether imide resin.
PEK	Polyether ketone resin.
PES	Polyether sulphone resin.
Phenolic	Thermosetting resin.
Polyimide	A thermoplastic or thermosetting resin with excellent high temperature properties also refered to as PI.
Ply	A single layer in a laminate.
Polyester	A thermosetting resin widely used with glass fibre.
PP	Polypropylene resin.
pph	Parts per hundred.
ppm	Parts per million.
PPM	Prepreg moulding.
PPO	Polyphenylene oxide resin.
PPS	Polyphenylene sulphide resin.
Preforming	A process by which fabric (or filaments) can be shaped into a desired form using a mould, before full impregnation – this is achieved by coating the reinforcement with a small amount of thermoplastic binder.
Prepreg	An intermediate fibre-reinforced plastic product which is ready for manufacture into a component – either in the form of sheet (for moulding) or tape (for winding).
Prepreg moulding	A process by which prepreg material is moulded either by autoclave or vacuum bag.
Press moulding	A process in which a press (cold or heated) is used to form an article from a compound comprising a mixture of fibres and resin.
PRM	Press moulding fabrication process.
PS	Polystryene resin.
PSP	Polystyryl pyridine, a thermosetting resin with excellent high temperature properties.
PSU	Polysulphone resin.
PUL	Pultrusion fabrication process.
Pultrusion	A process in which filaments and/or fabric coated with resin are pulled through a heated die and rapidly cured to retain the die shape.
PVC	Polyvinyl chloride resin.
R-glass fibre	A high strength version of E-glass.
RIM	Reaction injection moulding process.
RRIM	Reinforced reaction injection moulding.

Reaction injection moulding	A process in which the resin and its curing agents are rapidly mixed and injected into a mould; reinforcing fibres may also be added and then the product designated reinforced reaction injection moulding (RRIM).
Reinforcement	Fibres which have desirable properties which, when they are mixed together, enhance the properties of the host matrix.
Resin	The polymeric material in which the fibrous reinforcement is embedded.
Resin transfer moulding	A process in which the catalysed resin is transferred into a mould into which the reinforcement has already been laid.
Roving	A number of yarns, strands or tows which are collected into a parallel bundle with little or no twist.
RTM	Resin transfer moulding process.
Service temperature	Temperature at which a component is used in service.
S-glass fibre	A high strength version of E-glass.
Size	A treatment applied to yarns or fibres to protect their surface and facilitate handling.
SMC	Sheet moulding compound (see compound).
Specific property	Ratio of property to density or specific gravity.
Strand	An untwisted bundle of fibres.
Swirl mat	A type of mat in which a continuous fibre bundle is randomly laid on a plane.
Tex	A unit of linear density equal to the mass in grams of 1000 metres of filament or yarn.
T_g	Glass transition temperature of a resin.
Thermoplastic polymers	Plastics capable of being repeatedly softened by heat and hardened by cooling.
Thermosetting polymers	Plastics that once cured by chemical reactions or heat, become infusible solids.
Tow	A bundle of continuous filaments that are untwisted.
TSC	Thermoplastic sheet compound (see compound).
Twist	The number of spiral turns about the axis of a strand or yarn per unit length.
UD	Unidirectionally aligned fibres.
Vaccum bag moulding	A process in which a vacuum is used to to consolidate prepreg material which is laid up in a mould and enclosed in a vacuum tight bag.

Vinyl ester	A type of thermosetting resin with good mechanical properties.
Volume fraction	Fraction of fibres per unit volume of composite.
Weight fraction	Fraction of fibres per unit mass of composite.
Wet-out	Process by which the resin impregnates the yarns or fibres.
WR	Woven roving – a coarse woven fabric usually made from glass fibre tows.
XMC	Form of sheet moulding compound in which the fibre bundles are aligned (see compound).
Yarn	A bundle of filaments that have been twisted – generally used for processing into fabrics.

UNITS

property		units	designator
modulus	tensile	GPa	E
	flexural		
	compressive		
	tranverse		
	out-of-plane		
	shear		G
	torsional		
strength	tensile	MPa	σ
	flexural		
	compressive		
	transverse		
	out-of-plane		
	shear		τ
	torsional		
interlaminar shear strength		MPa	ILSS
strain		%	ε
elongation at break		%	ε
Poisson's ratio		–	ν
toughness (impact strength)		J/m or J/m^2	–
maximum fatigue stress/ ultimate stress		–	S
minimum/maximum fatigue stress		–	R
surface hardness (Barcol)		–	B
glass transition temperature		°C	T_g
heat distortion temperature		°C	HDT
thermal conductivity		W/m.K	CTC
thermal expansion coefficient		$°C^{-1}$	CTE
specific heat		J/kg.K	C_p
density		Mg/m^3	ρ
shrinkage (mould)		%	–
electrical resistivity		Ω m	–
dielectric constant		–	ε′
dissipation factor (Hz)		–	tan δ
flame spread (UL 94), see section 8.4		HB – 5V	–
oxygen index		%	–

property	units	designator
water absorption	%	–
viscosity	mPa.s	η
specific gravity	–	sg
specific modulus	GPa	E/sg
specific strength	MPa	σ/sg
fibre fraction – by volume	%	v/o
– by weight	%	w/o

RELATION OF WEIGHT TO VOLUME FRACTION

The graph shows how to convert from mass to volume fraction or vice versa for a fibre composite.

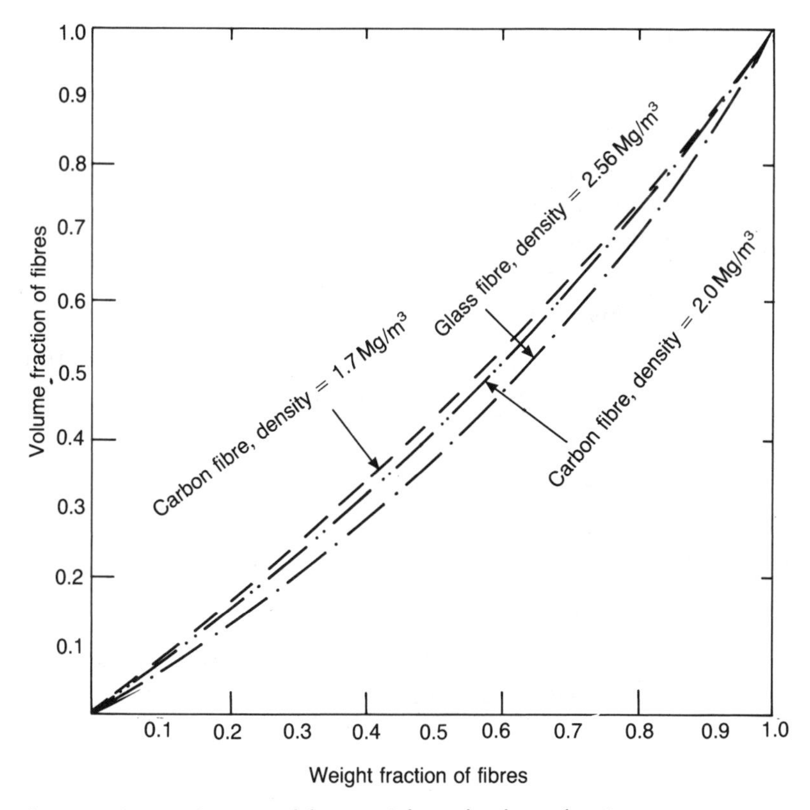

Figure 0 Conversion to and from weight and volume fractions.

Source: Design data for fibreglass composites (Owens Corning Fiberglas, Ascot).

FABRIC NOTATION

The following notation is used to designate fabrics in the absence of any agreed notation:

numbers	alignment angle;
(numbers)	mass in grams;
letters	method of fabrication;
SB	stitch bonded;
FW	filament winding;
W	woven;
I	inlaid;
K	knitted;
CSM	chopped strand mat;
+	combination fabric;
[number p]	number of plies;
/	separation of warp and weft rovings;
\	separation of plies.

Example 1: combination fabric

0/90	SB	(800)	+	CSM	(100)
alignment	process	mass	combination fabric	process	mass

Example 2: biased fabric

0	(567)	/ 90	(35)	W
alignment	mass	alignment	mass	process

Example 3: plied laminate

45	W	(250)	\ 0	(150)	[4p] \	45 W (250)
alignment	process	mass	alignment	mass	plies	

PLIED LAY-UPS

A common notation for specifying a laminate structure made from prepreg is:

$$[\pm\alpha_a \pm\beta_b \pm\gamma_c]_{ds},$$

where $\pm\alpha$, $\pm\beta$, $\pm\gamma$, etc. are the angles of the unidirectional fibre in relation to a reference axis; the integers a, b, c, indicate the number of layers; d the number of times the bracketed construction is repeated and *s*, if present, that the whole lay-up is symmetrical about a centre plane. Most constructions are symmetric so that distortion is avoided when the plate is cured. The + or − signs may not both be present, e.g.

$$[+45_2 0_4 90_2 - 30_3]_{2s}$$

represents a sequence of two +45° layers, four 0° layers, two 90° plies and three −30° plies with the whole sequence repeated twice and symmetrical about the centre plane.

1 DESIGN CONSIDERATIONS

1.1 INTRODUCTION

Designing with composites, as with any other material, is a combination of art and science. The creative element of the design process is often associated with shape, geometry, colour and surface texture, that is with the visual and tactile attributes of the structure or component. Beneath the surface are the technical aspects of the design; the choice of materials, their properties, loads, failure criteria, fabrication and processing methods.

Good design needs to integrate a number of elements which, all too often, are considered separately; size, geometry, loads and displacements, suitable materials, and so on. Lack of integration makes for an inferior and inefficient product. With traditional engineering materials, such as metals (and to some extent ceramics and polymers), it is possible to produce a working design, even if it is not an efficient one. This arises because, essentially, the properties of most of the materials are known to a high degree of accuracy, with relatively little ambiguity, and this information is available in reference books and is backed up by suitable design codes. The costs of traditional materials are also generally low, so that there may be no great benefit in excessive optimization. Another important consideration is that traditional materials are isotropic and essentially homogeneous; their mechanical and physical properties do not depend upon the direction in which they are measured and do not vary from point to point in a given volume, unless one considers a very small scale of structure. The only common materials which are different in these aspects are wood and concrete.

Fibrous composite materials are different, and it is essential to consider all the aspects of the design (including fabrication) together if an efficient, safe and cost-effective structure or artefact is to be produced. There are several reasons for this of which the designer must be aware:

- the composite material and structure are usually fabricated at the same time, and little or no modification to the final product is possible after manufacture;
- the materials are highly anisotropic and heterogeneous. This is a consequence of their fibrous structure and of the comparatively large size of the fibres (typically 10 μm diameter). The strength and stiffness properties of polymer composites reflect the properties of the fibres

only in the directions *parallel* to them; in any other direction they are, at best, a poor compromise of fibre and matrix properties and, at worst, the properties of the matrix alone (or the interface between fibre and matrix), which is by definition the more compliant and weaker component of the system;
- for any given combination of polymers and fibres there is a large number of possible materials which can be made, depending on the relative proportions of the components and the orientation of the fibres. Further variability is due to the effect of fibre length in short fibre composites, the mixing of fibres (hybrid materials) and the use of core materials to produce sandwich structures.

There are various design implications due to these differences:

- property data are relatively scarce and it is not always apparent how to use the information available to estimate the properties of other fibre/resin combinations;
- there is a need to understand how composite properties relate to those of the components;
- it is necessary to understand how composite properties change with fibre volume fraction and fibre orientation;
- care must be exercised to allow for the (sometimes) very low transverse and shear properties of the materials.

However, the very variety of raw materials and methods of combining them in the final product, plus the excellent specific properties (e.g. the ratio of the mechanical property to the specific gravity of the material), give the designer a new freedom to tailor the material to the function required.

In order to take full advantage of the potential of the fibre-reinforced polymer matrix composites in engineering applications, the (often additional) cost of materials and fabrication methods must be offset by design solutions which are directed to specific applications. Effective design with composites means exploiting all the advantages that they can offer to provide either competitive alternatives to existing materials or specific, unique, benefits such as mass reduction or a smaller number of parts. To achieve this it is important to think in terms of the composite from the beginning of the design process. One-for-one substitution of a metal component already integrated into a structure with a composite equivalent is unlikely to be satisfactory. The composite unit must not only be judged in terms of the cost of materials and fabrication but also in terms of, possibly, reduced weight or simpler assembly, hence less likelihood of malfunction, a longer life, a fail-safe failure mode, and so on.

A general consideration of designing with reinforced plastics is given by Mayer (1993).

1.2 LAYOUT OF THE BOOK

The theme of the twelve chapters which comprise this book is information which can be used for designing a fibre composite article or structure. The data cover a wide range. There is information on the short- and long-term thermomechanical and electrical properties and behaviour of fibre-reinforced polymer composites, plus their response to fire and environmental effects. Finally the influence of processing on properties and the quality assurance of the final product are also considered.

The majority of data refer to continuous, aligned, fibre systems since these have the highest performance. In practice, unidirectional plies or laminae are often stacked in a calculated sequence to produce a laminate with improved properties in the off-axis directions. However, the elastic properties of these constructions can be predicted using classical laminate theory, see Jones (1975) for instance, and it is possible to estimate the strength using a suitable failure criterion, provided in both cases the behaviour of the basic ply is known. Thus, in view of the multiplicity of possible laminate lay-ups, it appears more logical to concentrate on individual ply properties rather than on those of the more complex laminate. Most finished components are essentially planar, and through-thickness properties are assumed to be similar to transverse ones. There is evidence that in the future, composites with reinforcement in the through-thickness direction will become increasingly important in certain applications, but at present there is virtually no experimental information on properties in this direction.

The data included in this text on several specialized types of composite are rather limited. These (mainly thermosetting matrix) systems include those based on aligned short fibres, hybrids consisting of two or more types of fibre in a common matrix and sandwich structures consisting of a light-weight core and composite skins. These systems either have relatively low properties or such a range of possible constitutions as to make it difficult to find representative structures. Nevertheless, they are widely used in many areas, often because of their ease of fabrication.

Properties of typical thermosetting and thermoplastic polymer matrices and single fibres are listed. The former have relevance to the upper working temperature and off-axis properties of the composite, while the latter can be used to estimate composite strength and stiffness parallel to the fibre direction.

Finally, the really important topics of fabrication and quality assurance of the product are addressed. As was stated earlier, the composite materials and structures are often fabricated at the same time, so the fabrication process and the skill with which it is executed can make or break the design work. While this is true to some extent for all materials, it is much more important for composites. Allied to this is the topic of quality control. To obtain the optimum properties in a repeatable manner requires attention to, and close cooperation between, these two skills.

1.3 CHOICE OF DATA

Early studies of fibres and composite materials emphasized the tensile strength and modulus of the fibre, since this was the most convenient way of evaluating fibre properties and the flexural and tensile properties of simple unidirectional composites. Flexural measurements were particularly favoured as they were relatively simple to perform and provided modulus, strength and interlaminar shear strength (ILSS), the latter a measure of the fibre/resin bond. Critics pointed out that because of the nature of the bending of a beam, flexural modulus and strength were a mix of tensile and compressive properties. Even if failure occurred in an ILSS test, the shear strength obtained was related to the transverse compressive and longitudinal tensile strength as well as the properties of the matrix and fibre/matrix interface. In addition, it was shown that severe stress concentration could occur in the vicinity of the loading and support rollers. Another difference between tensile and flexural measurements is that a lesser volume of material is subjected to a maximum stress in the latter case. Hence the probability of encountering a critical flaw at a particular stress level is less, and so the flexural strength should be higher than the tensile value.

With improved fibre properties, fibre surface treatments were developed to aid bonding to the matrix, protect the fibre and improve processing. As the uses of composites expanded, two problems arose. First the question of test standards and second the extension of the database to include the considerable number of additional properties required by the designer.

The influence of test method is explored in one or two cases in Chapter 5 where it is shown, for example, that for compression strength the matter is still not resolved. Standard test methods for some fibre-reinforced plastics and high performance composite materials are given in Mayer (1993). The reader should be aware though that the issue of test methods is far from settled and research is still being vigorously pursued. Many companies, particularly in the aerospace industry, have their own well-established test procedures. Furthermore, as the properties of high performance carbon fibres are extended, established test methods may no longer be applicable.

It is very useful when assessing information to estimate the expected value, for longitudinal properties, from the rule of mixtures. This states that the composite property, X_c, is equal to the volume fraction weighted average of the contributions from the fibre and matrix, *viz.* $X_c = X_f V_f + X_m V_m$. As $X_f \gg X_m$, usually, the approximate form, $X_c = X_f V_f$ is often employed. The properties of the matrix are used to predict the transverse and shear performance of the composite. This approach should never be used as a substitute for accurate experimental data since many other factors are not taken into account in these simple approaches. Nevertheless it can alert the user to data that are too high for the raw materials used (i.e. in excess of the rule of mixtures), or excessively poor results that may point to a defect in the test method, fabrication route or efficiency with which the fibre or matrix property is taken up into the composite.

Fracture and impact properties should be specially mentioned since they are probably the subject of more argument than any other composite property. This arises because of the extreme complexity of composite failure under impact conditions. To use information generated on unidirectional specimens to predict the behaviour of laminates and structures is extremely difficult, if not impossible, but the reader should study Chapter 6 for more information.

It may be thought that after five decades of glass-reinforced plastics and nearly three decades of high performance fibre composite systems, a very large and extensive property database would be available. Unfortunately, this is not so for several reasons. First, there have been significant improvements in the types and properties of fibres, and to some extent resins, and in the bonding of the two, over the years. Thus, much early data may be atypical in that the raw materials were of inferior quality to those available today, and are possibly no longer available. Second, to characterize completely a composite system requires a lot of careful and expensive work, and this is rarely carried out, or if it is, made public. Finally, although a range of polymers is available, most fibres are used with a specific family of matrices (i.e. carbon and aramid fibres with epoxies, and glass fibres with polyesters and to a much lesser extent with phenolics or vinyl esters).

The data presented here have been selected with the aid of the RAPRA database on reinforced plastics and composites, from commercial trade literature and sources, from the composites literature and the authors' and their colleagues' private sources. We have always tried to select information on well-prepared and described systems using modern fibre types and matrices. Inevitably it has not always been possible to do this and in some cases we have had to use older data, information where the fibre or resin is not fully specified or material for other than unidirectional systems. In doing this we have had to balance the need for information with the uncertainties mentioned, but where we have done this we believe that the information quoted, though not ideal, is the best way of filling a real gap.

1.4 COMPOSITES BEHAVIOUR

There are certain aspects of composite behaviour which should be explained here. Hull (1981) can provide more detailed information.

Figure 1.1 shows two idealized load/deflection characteristics for fibres. In the case (a) the behaviour is linear to failure, while in (b) the fibre stiffens as it is extended. This type of behaviour is exhibited by some

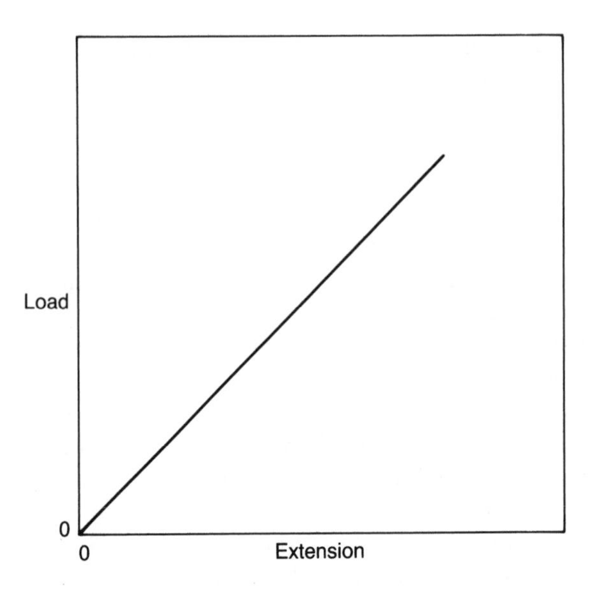

Figure 1.1(a) Idealized load/deflection characteristic for a fibre linear to failure.

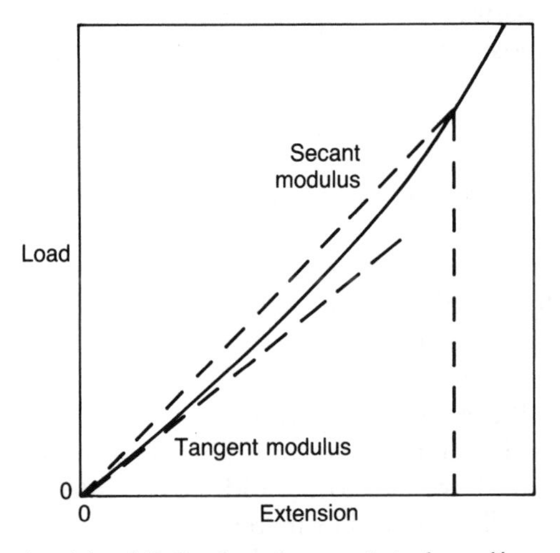

Figure 1.1(b) Idealized load/deflection characteristic for a fibre which behaves in a non-linear manner. Note definitions of tangent and secant modulus.

aramid and carbon fibres and is due to re-alignment in the internal structure of the fibre as it is stressed. The initial slope in the latter case defines the tangent modulus. The secant modulus, which is mentioned in connection with some fibre data in Chapters 3 and 5, is defined, for a specific strain or deflection, as shown in Figure 1.1(b).

Idealized load/deflection curves for unidirectional fibre composites are illustrated in Figure 1.2. In (a), the fibres all break at essentially the same strain. In (b), which is more typical of flexural stressing (hence the

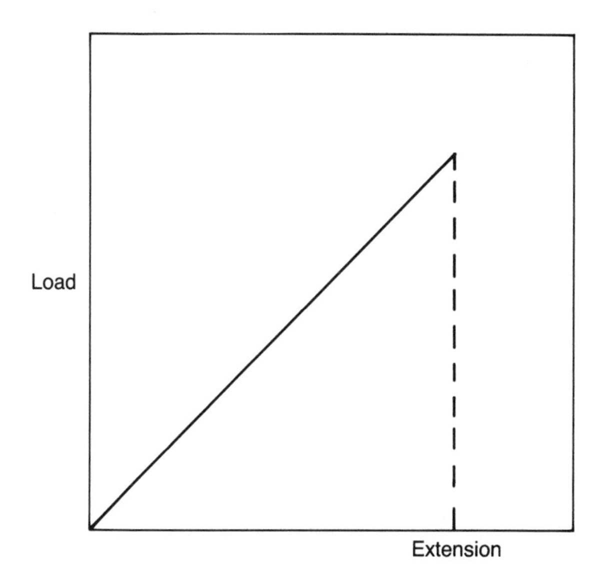

Figure 1.2(a) Idealized load/deflection curve for a unidirectional fibre composite in which all fibres fail at approximately the same strain.

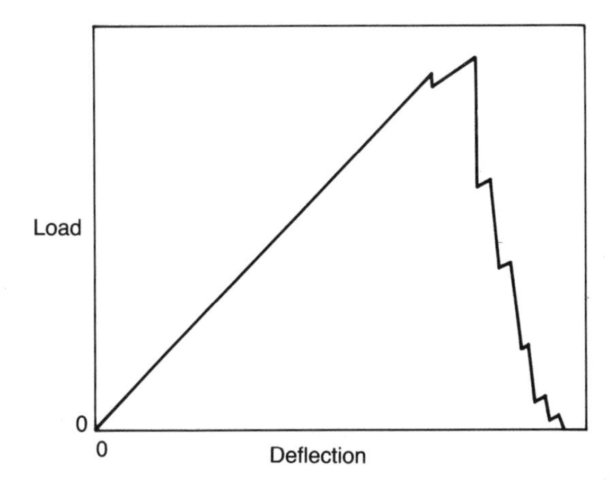

Figure 1.2(b) Idealized load/deflection curve for a unidirectional fibre composite in which failure occurs over a range of deflections – this may be associated with some flexural failures.

abscissa is labelled deflection), an initial load drop is apparent prior to the peak load. The majority of the fibres then break but because of differences in tensile and compressive behaviour in the composite, fibre pullout and other micromechanical mechanisms and fibres effectively blunting the failure cracks, the composite behaves in a tough manner and the load decreases slowly. The load deflection curves are not necessarily linear

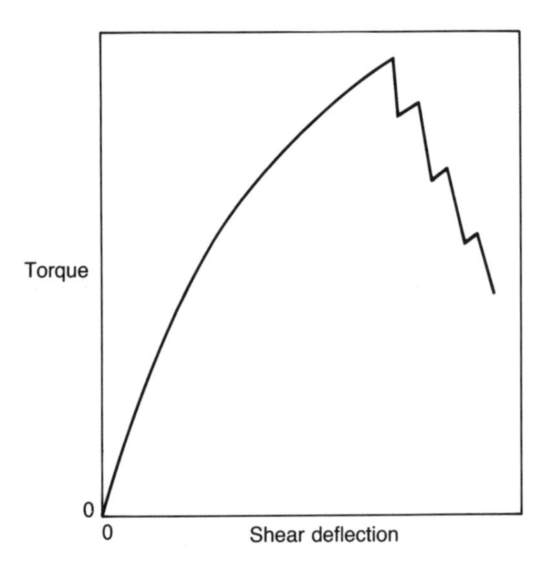

Figure 1.3(a) Torque/shear deflection curve for a well-bonded unidirectional composite.

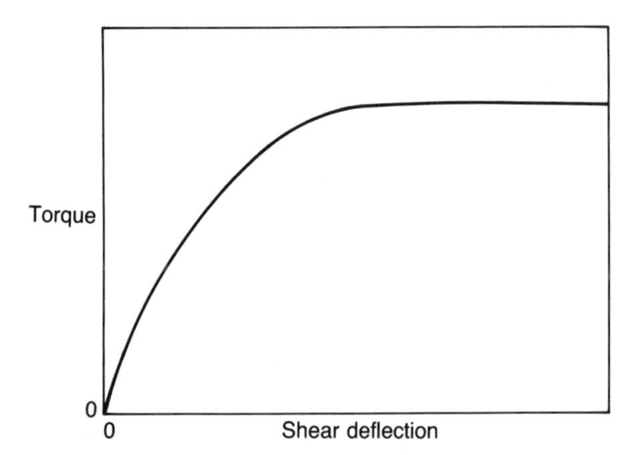

Figure 1.3(b) Torque/shear deflection curve for a unidirectional composite with a very poor bond between the fibre and matrix.

to failure. The modulus is generally defined as the initial or tangent modulus, and sometimes as the secant modulus (see Figure 1.1(b)).

Figure 1.3 shows two torsional, shear torque/shear deflection curves for unidirectional fibre composites. In (a), the fibre/resin bond is good and failure is initially due to that of the fibres in the most highly stressed outer region of the specimen, and continues as fibres nearer and nearer to the longitudinal axis of the specimen successively fail at a decreasing torque. In (b), debonding, but not fibre failure, occurs and the maximum torque remains constant with increasing shear strain.

Figures 1.4 and 1.5 illustrate the load/deflection behaviour of T300

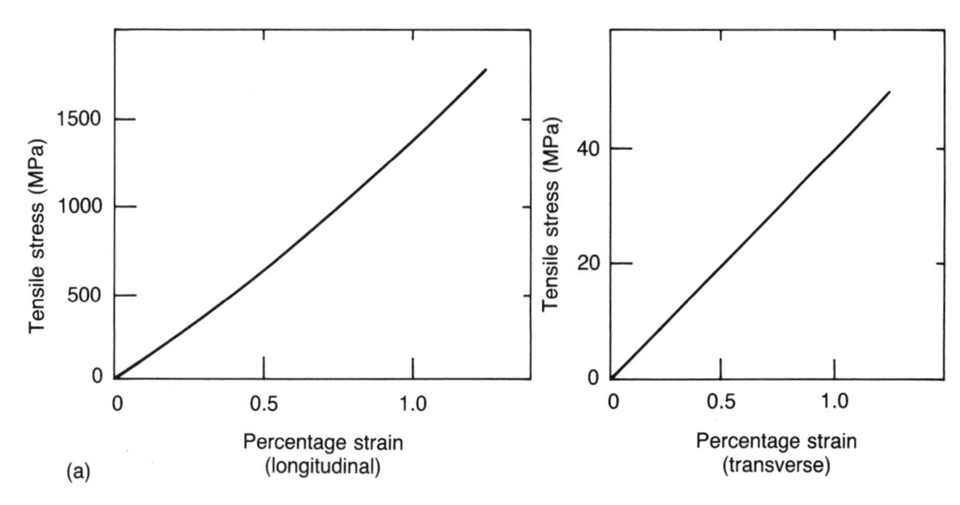

Figure 1.4(a) Longitudinal and transverse tensile load/deflection curves for a unidirectional carbon fibre composite (Hammond and Lee).

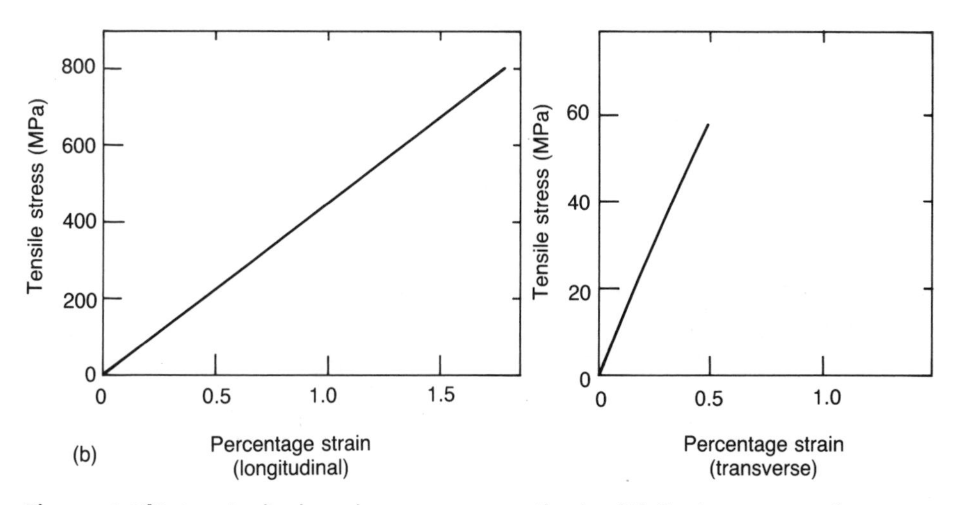

Figure 1.4(b) Longitudinal and transverse tensile load/deflection curves for a unidirectional glass fibre composite (Hammond and Lee).

carbon fibre epoxy resin and unidirectional E-glass specimens parallel and perpendicular to the long axis of the fibre, and similar information for balanced 0°/90° materials and a chopped strand mat specimen, respectively. The fibre volume loadings are 59 v/o and 48 v/o, respectively, in Figures 1.4(a) and 1.4(b), and 57 v/o and 53 v/o, respectively, in Figures 1.5(a) and 1.5(b). The loads at which the transverse ply failed (Figure 1.5b) or failure initially occurred (Figure 1.5c) are clearly marked by a change in slope, or 'knee'.

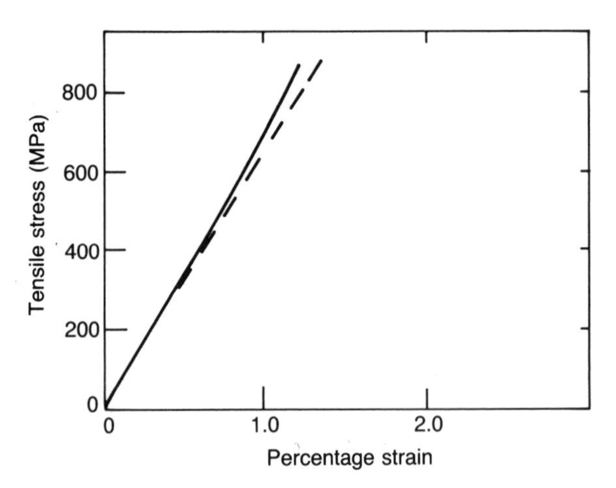

Figure 1.5(a) Load/deflection curve for a 0°/90° carbon fibre composite (Hammond and Lee).

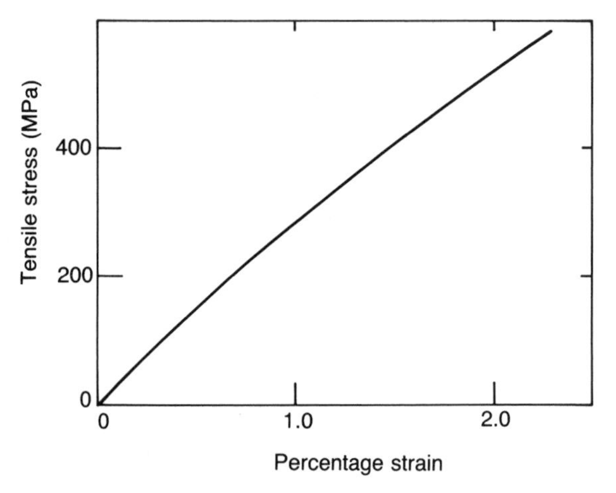

Figure 1.5(b) Load/deflection curve for a 0°/90° glass fibre composite (Hammond and Lee).

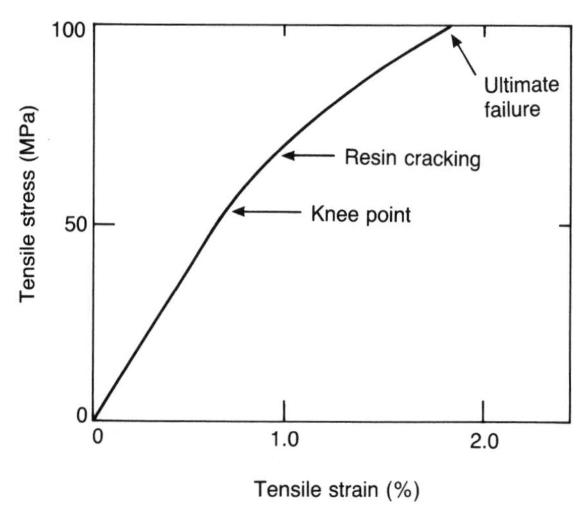

Figure 1.5(c) Tensile stress/strain curve for a chopped strand mat glass polyester laminate – note the knee (Mayer).

One very important difference between the behaviour both of fibres and composites on the one hand, and the more familiar metals on the other, is clear from Figures 1.1 to 1.5 – fibres and composites do not undergo true plastic deformation (notwithstanding the fact that thermoplastics may exhibit this type of deformation). In somewhat simplified terms, composites stressed in the fibre direction are elastic to failure. Even in Figure 1.3(b), the fibre to resin bond has been broken, and the two components are interacting through frictional forces. This difference in behaviour between fibre-reinforced materials and most metals is very significant. Despite the very great advantages enjoyed by composites due to their low densities (and hence very high specific properties), they do not, in general terms, possess the 'safety feature' of metals whereby localized overstressing or impact can be accommodated by localized plastic deformation and stress redistribution. Once a composite has been overstressed, albeit locally, permanent damage due to fibre failure, debonding or delamination is present and cannot usually be removed. This is clearly a very simplified treatment of a very complex problem but should nevertheless indicate an important difference between composites and metals.

Another important difference between the two materials is illustrated in Figure 1.6, where the tensile modulus and compressive strength of a unidirectional fibre composite are shown as a function of the angle of stressing to the longitudinal fibre axis. The very rapid fall-off in properties with off-axis loading is apparent. These figures are detailed in Chapter 5, where specific data are listed for longitudinal and transverse directions. To overcome the effects (although at the cost of a reduction in properties in the longitudinal direction) composites can be formed by stacking unidirectional plies in a specified sequence of orientations or by using woven fabrics. This behaviour illustrates the importance, when designing with composites, of knowing load paths accurately and ensuring that fibres are placed to take the loads.

One other topic must be mentioned briefly – reinforcement efficiency. It has been stated that it is possible to estimate (neglecting the resin contribution) longitudinal (i.e. fibre-controlled) properties on the basis of the fibre property and fibre volume fraction. This provides an upper limit to the composite property. The reinforcement efficiency is defined as the ratio of measured value for the composite (e.g. tensile strength) to that expected from the simple calculation. Values quoted in Chapter 5 indicate that in some cases this factor can be as little as 50% or less, and illustrate the importance of measured rather than calculated or estimated values of a property. The low reinforcement efficiency may be due to several causes including misaligned fibres, poor fibre/resin bonding due to incorrect surface treatment or fibre sizing or the use of an inappropriate matrix and so on.

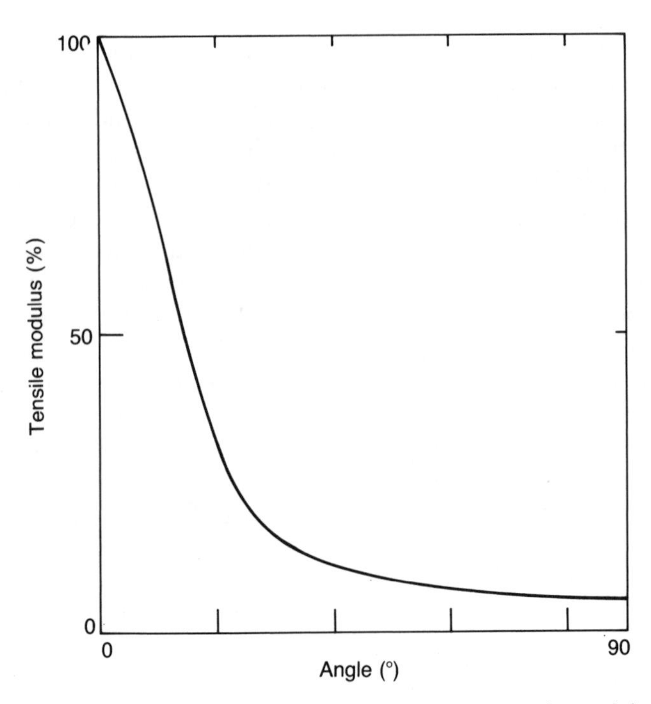

Figure 1.6(a) The effect of fibre orientation on composite tensile modulus. Calculated for a 60 v/o, unidirectional laminate. The angle refers to that between the fibres and the load axis.

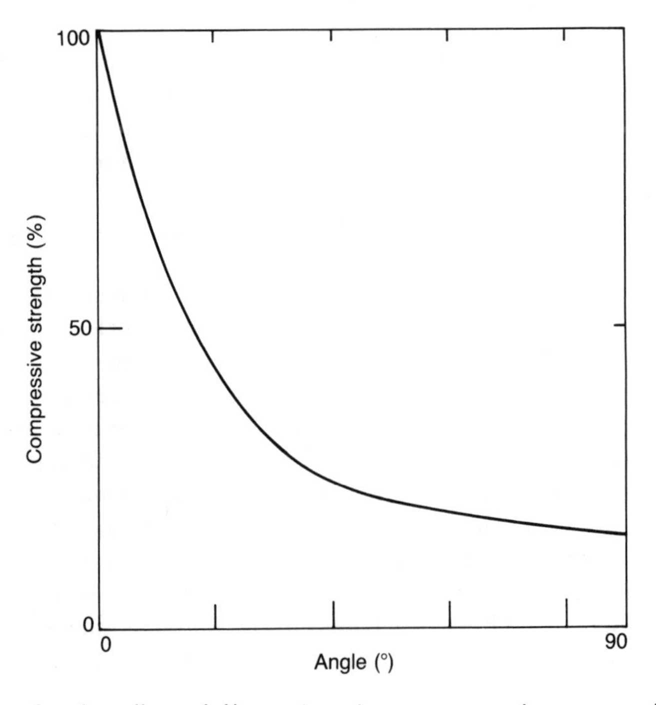

Figure 1.6(b) The effect of fibre orientation on composite compressive strength. Measured for a 60 v/o, carbon fibre laminate.

1.5 CONCLUSION

Design Data for Reinforced Plastics is what the title states – a compilation of data on the behaviour and performance of fibre-reinforced polymers. It is not meant to be read straight through at a sitting nor used without some understanding of composite materials and their own advantages and problems. However it does bring together in one place an extensive collection of all types of data relevant to using and designing with composite materials. Page locators have been provided to assist the reader in finding and following a particular topic or material and an extensive glossary is provided which covers the majority of technical terms and chemical abbreviations used.

In scanning so much material errors of understanding, interpretation and transposition may have occurred. The enquiring reader will, however, note other discrepancies if he or she delves into the composite literature to obtain further information on the specific properties of a particular system, or if he or she measures the properties of a material that they have bought or manufactured.

These arise because the properties of a composite depend upon the volume, disposition and nature of the fibres, the type of matrix, the bonding between the two components, the skill with which the artefact was made and how it was tested. Furthermore certain fibre types, especially carbon ones, have changed markedly over the years and the term epoxy or polyester covers a whole family of resins with a range of properties.

Although provided in good faith we are unable to accept responsibility for the accuracy of any of the information included in this guide or for the consequences that may arise from designs or constructions based on any of the information supplied or materials described. The inclusion or omission of a particular material in no way implies anything about its performance.

1.6 REFERENCE INFORMATION

Sources

Hull, D. (1981) *An Introduction to Composite Materials*, Cambridge University Press, Cambridge.

Jones, R.M. (1975) *Mechanics of Composite Materials*, McGraw-Hill/ Kogakusha Ltd, Tokyo.

Mayer, R.M. (ed.) (1993) *Design with Reinforced Plastics*, Design Council, London.

2 MATERIAL SELECTION AND DATA ASSESSMENT

SUMMARY

The major constituents that comprise a composite are considered in this chapter and a method of making an initial selection is outlined. The use of databases for material selection is briefly considered. The basis of collecting and assessing data for inclusion in this book is discussed.

2.1 INTRODUCTION

The material selection process can be complex in the sense that a wide variety of constituents is available, and there are few standard materials in use. This chapter outlines a coarse method of selection and shows how databases allow a more refined choice to be made.

In the next chapter, properties of the fibres and resins are described in more detail, so allowing a more quantitative choice to be made.

In general, selection may be undertaken either from a design value for a specific property which must be associated with a suitable material combination or a material combination for which an appropriate property level is sought.

Within this reference frame, the process of selection is outlined. For many material combinations one finds that there may be limited data available and only for the most commonly used materials is it possible to compare data from different sources.

Finally, some comments are included on the self-consistency of data and how to interpolate or extrapolate in a safe manner.

2.2 MATERIAL CONSTITUENTS

The principal constituents of a fibre-reinforced plastic are listed in Table 2.1.

The roles of the reinforcement, resin and core material have already been considered (Chapter 1). The other constituents play a lesser, but still important role, in formulating the properties of a composite.

Fibre size and finish

Most fibres are coated with a chemical size to assist the wet-out and bonding between the resin and the fibre. Fibre bundles are coated with an organic finish to facilitate handling during processing and moulding.

Table 2.1 Principal constituents

fibre
resin
core material
fibre size and finish
additives

Size
Size is not generally specified other than the percentage that is applied and the generic type. Size is an organic compound which is chosen to be compatible with the resin with which the fibre is used – generically denoted as silane, chromesilane or epoxy, of which silane is the most common for polyester resins.

Finish
Finish is a mixture of compounds that will generally depend upon the handling that the fibre bundles will be subjected to during processing. For weaving, for example, a higher level is applied, whereas for high speed wet-out, a smaller quantity is applied.

To obtain a high level of properties, it is essential that a suitable size and level of finish be selected. The fibre and resin manufacturer should both be consulted if a combination is selected which is unfamiliar.

Additives

Additives are taken to include substances which can polymerize the resin or improve the properties of the composite.

Polymerization additives
These include the catalyst or a hardner and possibly an initiator or accelerator. They are not listed in this data set unless they form an important part of the resin's composition such as may occur with epoxy resins. Otherwise it is assumed that the moulder will use the appropriate type and quantity, in accordance with the supplier's instructions. In addition, it is assumed that the composite is fully cured; this might well involve applying a temperature cycle after curing (section 12.3).

Other additives
These range from plasticizers and flame retardants to inert fillers. These are also not listed as they form the basis of resin blends. They are able to affect many properties such as elongation, flammability and viscosity (Table 2.6).

2.3 MATERIAL SELECTION

The selection criteria in this data set are based on their common usage for general engineering applications. Certain sectors like aerospace and the chemical industry use a restricted range of materials, which are fully characterized and are detailed elsewhere.

The data set is further reduced by the availability of published data. The scan of the RAPRA bibliographic base reveals that there is a close correlation between material usage and the amount of published data. Thus there is much information available on glass fibre/polyester, which has been in existence for fifty years, and little on carbon fibre/PEEK, which has only been established during the 1980s.

This trend is reflected in our choice of materials and data.

Constituent choice

Reinforcement and resin

A simple ranking system is used to establish an initial choice (Tables 2.2, 2.4). This provides a way of selecting a material based on a critical design requirement; however the level of the other properties or parameters will also need to be checked by using the data set that follows or a database. As the design proceeds, the initial choice may need to be reassessed.

It is worthy of note that the reinforcement and resin can generally be chosen independently so that this should permit two property levels to be attained.

Fillers

These are selected primarily for their ability to impart specific properties, and these are generally incorporated by the manufacturer in response to market needs (Table 2.6).

Compounds

Compounds are intimate mixtures of reinforcement, resin, and fillers blended together for a particular manufacturing process. Within this group are included (Table 2.7):

- short fibres in a thermosetting resin like sheet moulding compounds (SMC), bulk (BMC) or dough moulding compounds (DMC);
- long fibres in thermosetting resin ('prepregs');
- long fibres in thermoplastic resin (TSC or 'thermoplastic sheet compound');
- random mat in thermoplastic resin (GMT or 'glass mat thermoplastics').

Their generic characteristics are set out in Table 2.8.

Core materials

These are selected by the designer on the basis of their properties (Table 2.9). They are primarily used to reduce mass and cost and to increase stiffness. They are rarely used in tension.

■ 2.3 REINFORCING FIBRES

■ **Properties**

Reinforcing fibres

The principal types of fibres are set out together with a property ranking of low to high (Table 2.2).

Observations

Fibres can be grouped according to their properties and cost. Glass, being the most popular reinforcement, has the largest number of sizings for use with different resin systems and processes. Polyester and cotton yarns are generally used together with other reinforcements in making fabrics by inlaying or stitch bonding.

Design implications

Each fibre type has a range of derivatives giving rise to a range of mechanical properties and costs. With glass it is possible to vary the chemical composition to produce a variety of fibres for different applications (Table 2.3). For carbon and aramid fibres, a wide variety of fibre types exist and these are described in some detail in Chapter 3.

Table 2.2 Fibre selection by property level

	strength	modulus	ease of bonding	abrasion resistance	flame spread	breaking strain	cost
E-glass	**	**	***	***	**	**	low
R-glass	***	**	**	***	**	**	med
aramid (HM)	***	**	*	*	***	**	high
carbon (HS)	***	***	***	*	***	*	high
polyethylene	**	**	*	*	***	***	high
polyester	**	*	*	*	***	***	low
polyamide	**	*	**	**	***	***	low
cotton	*	*	***	*	**	***	low

Notes: Ease of bonding – availability of sizes.
Abrasion – resistance to wear (section 9.4).
Flammability – UL 94 rating (section 8.2).
Cost (per kg) (1992 bulk prices) – low <£1.25
– medium £1.25–£10
– high >£10
Ranking * – low
** – medium
*** – high
HM – high modulus version of aramid fibre; HS – high strength version of carbon fibre (Table 3.1).

Types of glass fibres

A variety of fibres of varying composition has been manufactured for various applications (Table 2.3). E-glass is the most common and all results refer to this material unless otherwise indicated.

Table 2.3 Types of glass fibre

E	standard reinforcement, low alkali content <1%
A	high alkali content (10–15%); inferior properties to E; not widely used
C	improved corrosion resistance over E; usually used in form of surface tissue
E-CR	boron-free; good acid corrosion resistance; mechanical properties similar to E-glass
D	high silica and boron content; dielectric applications; radio frequency transparent
R, S-2	better mechanical properties than E-glass; limited to specialist applications due to higher cost
AR	alkali-resistant glass fibres for reinforcement of cement

Sources

Lovell, D.R. (1991) *Carbon and High Performance Fibres Directory*, 5th Edition, Chapman & Hall, London.

Thomas, E.J. *Coated Polymers in Defence*.

■ 2.3 RESIN MATRICES

■ **Properties**

Resin matrices

The principal resins are set out in Table 2.4 together with a property ranking of low to high.

Observations

A wide range of properties is available, cost also varies widely and depends markedly on the quantity bought. Prices in Table 2.5 are for bulk purchases and it should be noted that the cost of buying small quantities is higher. Many blends are available to extend the basic property range.

Heat is necessary to shape a thermoplastic, but not cure it. Heat speeds up the cross-linking of a thermoset and may be necessary for a full cure (Table 9.7 shows how the resin's HDT is increased by post curing).

Design implications

If one property is vital to the design then the resin should be selected to meet this criterion; otherwise the designer should aim for the appropriate balance of properties.

Table 2.4 Resin selection by property

	tensile strength	high temperature capability	flame spread	chemical resistance	mould shrinkage	cost
thermosets						
polyurethane	***	*	**	***	med	low
methacrylate	**	*	**	–	high	–
polyester	**	**	***	**	high	low
epoxy	**	**	**	**	low	med
vinyl ester	**	*	**	***	low	med
phenolic	*	***	**	**	low	low
polyimide	***	***	**	**	low	high
bismaleimide	***	***	**	**	low	high
polystyryl-pyridine	**	***	–	–	–	high
thermoplastics						
PP	*	*	***	***	high	low
PA	*	**	***	***	high	low
PES	*	***	**	***	med	med
PEI	**	**	**	***	med	med
PSU	*	**	**	***	med	low
PAS	*	***	**	***	med	–
PAI	***	***	**	***	med	high
PPS	**	***	*	***	med	med
PEEK	**	**	**	***	high	high

Notes: Rating values are given in Table 2.5.
See glossary for full chemical names and common tradenames.

Table 2.5 Rating values for resins in Table 2.4

property	rating	thermosets	thermoplastics
tensile strength (MPa)	*	<50	30–90
	**	50–70	90–140
	***	>70	>140
high temperature capability (°C)	*	<120	
	**	120–180	
	***	180–300	
flame spread UL 94 (see section 8.4)	*	5V	
	**	V-0, V-1, V-2	
	***	HB	
chemical resistance (to weak alkali)	*	poor	
	**	fair	
	***	good	
mould shrinkage (%)	low	<0.5	<0.5
	med	0.5–1.0	0.5–1.0
	high	>1.0	1.0–1.5
cost (£/kg) (1992 prices)	low	<0.8	
	med	0.8–2.0	
	high	>2.0	

More process options are available for thermosetting composites than thermoplastic composites at present, but the latter range is gradually being extended (Chapter 11).

Sources

Plascams: Plastics, a Computer Aided Materials Selector (1991) RAPRA, Shawbury.

Rubin, I. (ed.) (1990) *Handbook of Plastic Materials and Technology*, Wiley-Interscience, New York.

■ 2.3 FILLERS

■ **Properties**

Filler and other additives

The principal fillers are set out in Table 2.6 together with the major reasons for their use.

Observations

A variety of properties can be attained. Blending is usually done by the resin supplier, but additional blending can also be done at the moulding stage if convenient. Processability, flame retardancy, control of shrinkage and pigmentation are the principal reasons for using fillers. A brief description of the principal fillers follows:

- *Aluminium trihydrate* is typical of a range of materials that contain water of hydration – this is released at the flame temperature. This method does not produce any known toxic by-product.
- *Calcium carbonate* is an extensively used filler due to its low cost; commonly used in moulding compounds.
- *Carbon black* is widely used as a pigment; it also provides self-lubricating properties and increases conductivity (electrical and thermal).

Table 2.6 Typical fillers and principal reasons for their use

	extender	colour	mechanical properties	shrinkage reduction	key property improvement
aluminium trihydrate				•	flame retardancy
calcium carbonate	•	•	•	•	UV stabilizer
carbon black	•	•	•	•	conductivity
clay (kaolin)	•		•	•	mould flow
glass spheres (hollow)	•		•		density reduction
metallic oxides	•	•		•	conductivity
mica flakes	•		•	•	dielectric strength and heat resistance
rubber particles			•		toughness
silica	•		•	•	thixotropy
talc	•	•		•	cost reduction
wood flour	•			•	cost reduction

Notes: Fillers are in powder form unless indicated.
Extender – a bulk filler.
Colour – ability to pigment.

- *China clay (kaolin)* gives improved flow in moulding processes and hence ensures uniform distribution of reinforcement. It also improves chemical resistance, hardness and electrical resistivity. It is widely used in compounds.
- *Glass microspheres* due to their spherical nature can be easily moulded as polymer can flow with ease around spheres. Also impart toughness. Have a typical diameter range of 20–200 μm, and can be either solid or hollow.
- *Metallic oxides* and powders are used to enhance electrical and thermal properties.
- *Mica flakes* have excellent dielectric properties, electrical and arc resistance. Increase stiffness and dimensional stability.
- *Rubber particles* provide a means of toughening resins by the presence of small spheres of elastomeric material. Typical diameter range between 0.5 and 5 μm.
- *Silica* is used both as an extender and as a means of promoting thixotropy.
- *Talc* platelets bond strongly to a resin matrix thus improving some of the mechanical properties such as stiffness.
- *Wood flour* is widely used as a filler in thermosetting resins.

Design implications

Fillers can alter various properties and so extend the range of applications of both resins and composites by either altering the properties and/or processability – for example adding a filler to reduce mould shrinkage, may also increase the modulus.

Sources

Richardson, T. (1987) *Composites: A Design Guide*, Industrial Press, New York.

Rubin, I. (ed.) (1990) *Handbook of Plastic Materials and Technology*, Wiley-Interscience, New York.

■ 2.3 FILLERS

■ **Shrinkage**

Flexural modulus

Shrinkage and flexural modulus

The effects of fillers on these two properties are illustrated in Figures 2.1 and 2.2.

Shrinkage is important in processing as one normally moulds direct to final shape. Apart from using a low shrinkage resin (Table 2.4), the other option is to use a filled resin. Adding filler tends to increase the viscosity, so there is a trade-off between these two processing parameters.

The stiffness shows a linear increase as more calcium carbonate filler is added to polypropylene (Figure 2.2). The effect of the particle shape is demonstrated by glass fibres imparting greater stiffness to the polymer than the flakes of talc or the cubed crystals of calcite.

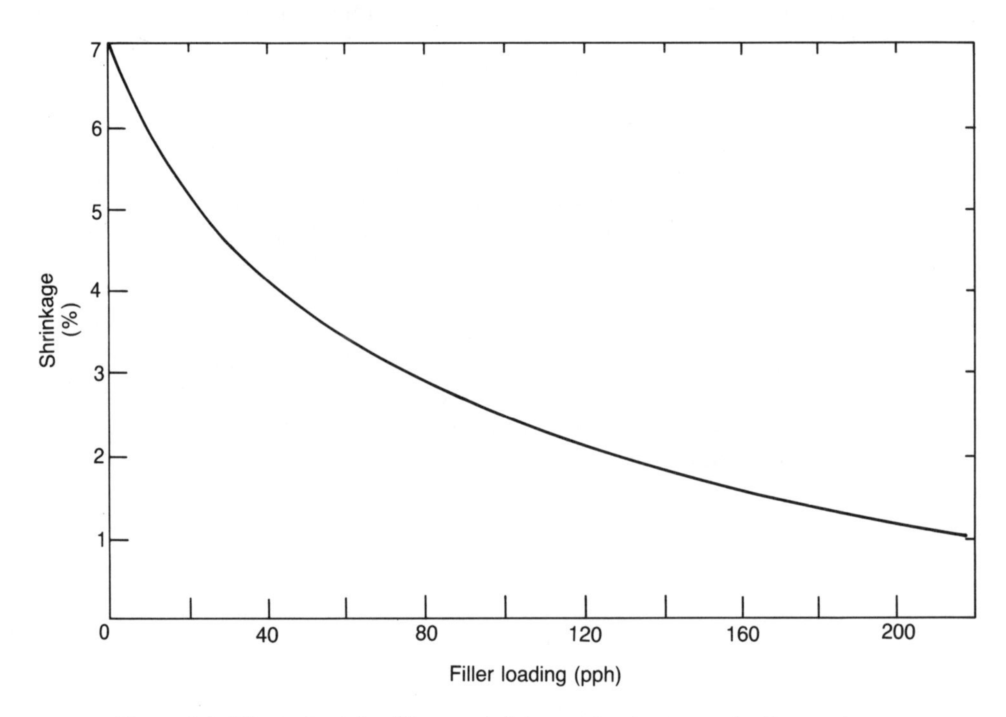

Figure 2.1 Effect of calcite filler on shrinkage of polyester resin (Pearson).

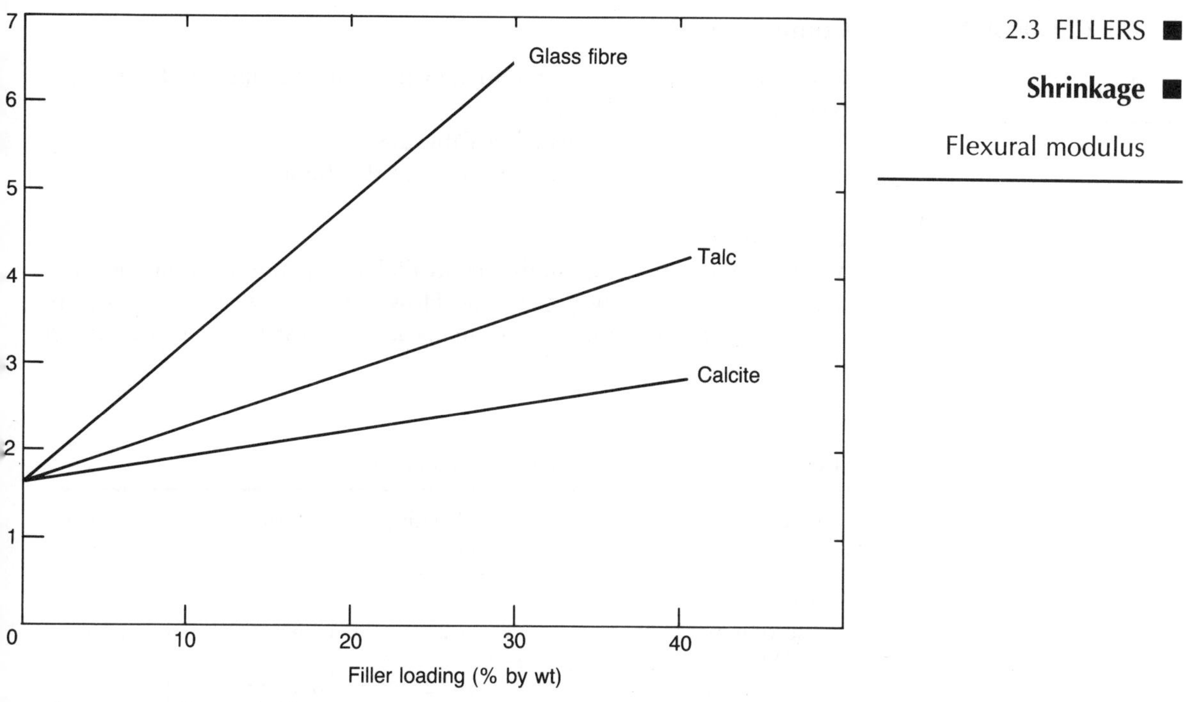

Figure 2.2 Effect of filler particle shape on flexural modulus of polypropylene (Pearson).

Sources

Pearson, P.G. (1988) Filled thermosets and thermoplastics, *Proc. Reinforced Injection Processing*, Paper 4, PRI, London.

2.3 COMPOUNDS

Properties

Compounds

Compounds are a blend of reinforcement, resin and filler, and the major types are listed in Table 2.7.

Typical properties are listed in Table 2.8 except for injection moulding compounds, which are not considered in this book.

Observations

Various types of compounds are available depending upon the fibre length, form and moulding process. However, most sheets and tapes are fairly thin (0.2–0.3 mm) and so many layers must be used to build up thick sections.

Table 2.7 Major forms of moulding compounds

compound	fibre length (mm)	form	moulding process
thermosets			
injection moulding (IMC)	1–2	compound	injection
dough moulding (DMC)	3–12	compound	press
sheet moulding (SMC)	20–25	sheet	press
(XMC)	continuous	sheet	press
prepreg	continuous	sheet	prepreg
	continuous	tape	tape winding
thermoplastics			
glass mat thermoplastic (GMT)	random mat	sheet	press
thermoplastic sheet compound (TSC)	woven roving	sheet	press
	aligned fibres	tape	press

Notes: For moulding processes, refer to Table 11.1.
Terminology is not always consistent in the literature.

Table 2.8 Properties of moulding compounds

	fibre w/o (%)	tensile strength (MPa)	tensile modulus (GPa)
SMC	30	85	13
DMC	20	45	9
GMT	30	85	6
TSC*	70	910	50
XMC*	70	485	35

Notes: * Essentially unidirectionally reinforced;
GMT – glass mat/PPO;
TSC – aligned fibre/PPS.

Design implications

Both the material constituents and the process route have been optimized by the material supplier, which permits reliable properties with small scatter.

Other than prepregs, most formulations are generally used for large batch or volume production using appropriate tooling and plant. Such moulding processes have limitations on both size and shape. Prepregs can be readily fabricated, even on a large scale, by vacuum bagging (section 11.5).

Source

Mayer, R.M. (ed.) (1993) *Design with Reinforced Plastics*, Design Council, London.

■ 2.3 FOAMS AND OTHER CORE MATERIALS

■ **Properties**

Foams and other core materials

Core materials range from polymeric foams of both thermosetting and thermoplastic type to a light wood like balsa, Table 2.9.

Observations

Core materials are characterized by their low density compared with either the resin or the fibre. Whilst some can be used separately as structural foams, these materials are often used as a core sandwiched between two laminates.

Of the thermosetting foams, both polyurethane and phenolic are available in a variety of densities and forms, with a wide variety of properties. Coremat is a flexible, lightweight, polymeric material, which is available in sheet form. Balsa wood is also light and stiff, and available in various thicknesses.

Design implications

Core materials are generally used where thick sections are required, but mass has to be minimized. In the aircraft industry, honeycomb core is used both to reduce mass and cost, but care must be taken to avoid moisture pick-up.

Table 2.9 Representative properties of core materials

	density (Mg/m^3)	HDT (°C)	compressive strength (MPa)	oxygen index (%)	water absorption (%)	cost
thermosets						
polyurethane	0.6	65	69	24	0.2	medium
urea-formaldehyde	0.01	75		30	20	low
phenolic	0.08	90		40	15	high
thermoplastics						
PP	0.6	46	14	17	0.02	low
PVC	0.7	55		40	0.2	low
PS	0.08	70	0.46	18	0.07	low
ABS	0.85	82	33	19	0.6	medium
PC	0.85	125	35	25	0.18	high
PES	0.9	>180	79			high
others						
coremat	0.6		22			low
balsa wood	0.1–0.25		1.2–3.6		significant	low
honeycomb	0.03–0.15	150–200	1–12		low	high

Notes: Costs (1992 prices) low – <£1.20/kg; medium – £1.20–£2.00/kg; high – >£2.00/kg.

Sources

Ashby, M.F. and Waterman, N.A. (1991) *Material Selection Guide, Volume 3: Plastics and Composites*, Elsevier, Barking.
McDonald, D. (1988) Resins, cores and fillers, *Proc. British Plastics Federation Conference*, Blackpool.
Plascams: Plastics, a Computer Aided Materials Selector (1991) RAPRA, Shawbury.

2.3 FOAMS AND OTHER CORE MATERIALS

Properties

2.4 DATA ASSESSMENT

Data have been obtained from two prime sources – published literature and the material suppliers.

Published literature has been primarily accessed via RAPRA's bibliographic base using a suitable search strategy. This has been supplemented from other sources including references cited in the original source documents. Where possible, additional information has been sought directly from the material suppliers.

The time frame is primarily from 1985 onwards for the following reasons:

- fibre sizes are continually changing as the technology advances;
- RAPRA's inclusion within their base of a descriptor which confirms the presence of numeric data;
- better agreement about test procedures.

Materials suppliers have a vast amount of information, mainly unpublished, which has been accumulated in characterizing their own products singly and in combination with other resins or reinforcements.

This information has been used selectively to infill gaps that appear in the literature and to provide the basis for inter-comparison with other data.

The assessment strategy is listed in Table 2.10.

Table 2.10 Data assessment strategy

material specification
fabrication procedure
test methodology
reliability
scatter
reproducibility
fit with other data

Data were considered in terms of the material constituents, specification and fabrication method and whether these were typical of the resulting composite. It has been tacitly assumed that authors have used valid test methods even though these have not always been cited. Reliability, scatter and reproducibility were difficult to assess unless the authors had commented on their data. Comparison with other data has generally been undertaken with the aid of the rule of mixtures for calculating the modulus (section 1.3).

Unfortunately not all data cited in this compendium meets these criteria and this seems to arise from the purpose for which the data were collected.

What therefore appears is a data set based on the reported information and a judgement by the authors as to the suitability of the data based on Table 2.10.

2.5 TEST METHODS AND STANDARDS

There are many methods and standards and these are being refined as markets expand. The formation of the single European market is providing a strong commercial incentive to reach agreement at an international level both within the market and with other trading areas.

In compiling the data set, test methods have been used as an indication of the way in which the data were collected. They are quoted without comment favouring one or other method. The various benchmark tests now under way will help to provide better correlations between various data sets.

2.6 DATA REDUCTION AND PRESENTATION

Data have been reduced where information is available from more than one source or where it is more meaningful to represent the data in a different way from that used by the original author. The reader should consult the source document if more detailed information is required.

Where scatter about a mean is reported, this is incorporated by scatter bars on graphs or directly within the tables.

Data points are shown on graphs where it is likely that the designer would wish to interpolate or extrapolate. In undertaking this type of approach the designer should check directly the critical part of the data or at very least confirm the trends that are being surmised. One extrapolation that is permitted within the codes is that of GRP pressure pipe where BS 5480 allows an extrapolation to be made (Figure 2.3).

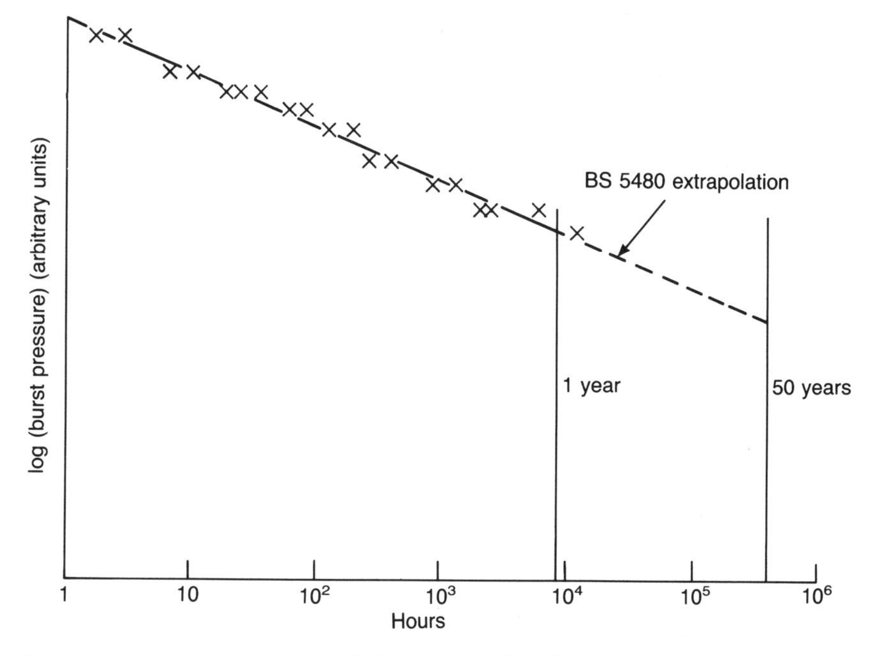

Figure 2.3 Burst pressure extrapolation permitted by BS 5480 (Greatorex).

2.7 USE OF DATA IN DESIGN

The data presented here can be used for establishing both the conceptual and embodiment designs. However to detail the design, data should be obtained from test pieces manufactured by the *same* process route as the component or structure to check the material combination, process and process route. This step should never be omitted.

Data collected from coupons cut from test plates may not be characteristic of the component as the fibre orientation, length and volume fraction may differ. Either the fibre orientation has to be made closer to that of the real component or else the design has to be altered at the detailing stage to account for the variations in properties from those assumed.

Use of databases

A database, like RAPRA's Plascams, enables the material choice to be refined more closely. Both properties and design parameters like cost may be specified and ranked (Table 2.11).

As with Tables 2.3 to 2.7, this choice needs to be checked with other data in the database or in this book.

Table 2.11 Material selection and ranking for a glass-reinforced composite

	flexural modulus	heat distortion	oxygen index
bisphenol	8	8	4
phenolic	7	7	7
epoxy	8	8	5
polyimide	8	9	8

Note: Ranking – 1 (poor) to 9 (excellent).

Sources

Greatorex, C.B. (1987) GRP in pressure pipelines. *Civil Engineering*, September 1987.

Plascams: Plastics, a Computer Aided Materials Selector (1991) RAPRA, Shawbury.

2.8 REFERENCE INFORMATION

General references

Johnston, A.F. (1986) *Engineering Design Properties of GRP*, BPF, London – useful source of information though compiled in 1978.

Engineered Materials Handbook, Volume 1: Composites; Volume 2: Engineering Plastics (1987) ASM International, Metals Park, Ohio – comprehensive compilation of data with explanatory text.

Richardson, T. (1987) *Composites: A Design Guide*, Industrial Press, New York – good summary of fillers and their properties.

Zweben, C. *et al.* (1989) *Mechanical Behaviour and Properties of Composite Materials*, Technomic, Lancaster – overview of mechanical properties with theoretical explanations.

Material constituents

Crystic Polyester Handbook (1990) Scott-Bader, Woolaston – useful information on curing and post-curing of composites.

Materials selection

Ashby, M.F. and Waterman, N.A. (1991) *Material Selection Guide, Volume 3: Plastics and Composites*, Elsevier, Barking – very extensive data.

ESA Handbook on Composites (1989) ESA, Nordwijk – comprehensive compilation of data for space applications.

Chemres II: Plastics and Rubbers Computer Aided Chemical Resistance Selector (1991) RAPRA, Shawbury.

Plascams: Plastics, a Computer Aided Materials Selector (1991) RAPRA, Shawbury.

Peritus: Engineering Materials Database (1991) Matsel Systems, Liverpool.

The three databases above are designed to select optimum materials for specific requirements; some data is available in them for each material, but is not comprehensive.

Data assessment

Bibliographic base for plastics, 1991, RAPRA, Shawbury – some 450,000 entries dating back to 1917.

Test methods and standards

Mayer, R.M. (ed.) (1993) *Design with Reinforced Plastics*, Design Council, London – Chapter 7 surveys existing test methods.

Perinorm Database (1993) BSI, Milton Keynes – complete set of standards for UK, France, Germany and international standards ISO, IEC and CEN/CENELEC.

BSI Standards Catalogue (1993) BSI, Milton Keynes – current list of UK standards.

ASTM Standards Handbook, Volume 8: Plastics (1991) ASTM, Philadelphia.

Benchmark tests: VAMAS – international collaboration on materials characterization and testing. Contact Graham Sims, NPL, Queens Road, Teddington, Middlesex TW11 0LW, UK, Tel. (44) 081 977 3222.

Data reduction

BS 5480: Specification for GRP pipes, joints and fittings for use for water supply and sewerage (1991) BSI, Milton Keynes.

Data in design

BS 7000: A guide to managing product design (1989) BSI, Milton Keynes. Check list of design tasks – design stages are described more fully in Mayer, 1993, Chapters 1 to 3.

Information sources

RAPRA Technology Ltd, Shawbury, Shrewsbury, Shropshire, SY4 4NR, UK, Tel. (44) 0939 25 03 83. Supplier of bibliographic, Plascams and Chemres databases – also provides general information.

Materials Information Service, Design Council, 28 Haymarket, London, SW1Y 4SO, UK, Tel. (44) 071 839 80 00. Helpline, information service, access to databases.

Institute of Mechanical Engineering, 1 Birdcage Walk, London, SW1H 9JJ, UK, Tel. (44) 071 222 78 99. Access to various bibliographic bases and Perinorm.

Databases

These are becoming more readily available for PC-compatible computers – the larger bases like Perinorm and RAPRA bibliographic base require a CD-ROM reader – smaller bases like Matsel's Peritus and RAPRA's Plascams and Chemres are available on floppy disks.

Search strategies

Databases contain a vast amount of information and each requires its own search strategy to obtain effective access to the information.

For this book, a set of descriptors was devised with the aid of RAPRA to scan the entries in their bibliographic base; for example glass-reinforced epoxy had 947 entries of which 203 are books and conference proceedings and 40 are monographs, trade references and standards.

Adding the descriptor 'hard data' reduced the entries to 200 and these were further reduced by adding descriptors like flexural properties and fatigue properties. Readers could use similar strategies to obtain specific information for their application.

3 PROPERTIES

SUMMARY

To use composites it is helpful to understand a little about their background and how, for instance, composite properties reflect those of the fibre and matrix. This information plus some fibre and matrix properties and comments on the physical forms of the materials and design and fabrication, is given here. This chapter augments the previous one and emphasizes different aspects of composites technology.

3.1 INTRODUCTION

To use the data in this book effectively and with confidence it is helpful to have an understanding of composites and how they differ from the conventional engineering materials with which we are more familiar.

The appeal of continuous fibre-reinforced polymers, or composite materials, is based on their properties, on the freedom they give the designer to tailor these properties to suit a specific situation and the relative ease with which composites can be fabricated.

The performance of the composite depends on that of the reinforcing fibre, the binding matrix and how these are bonded together. Individual fibres have a high modulus, from approximately 70 GPa to over 800 GPa, and strengths from approximately 1 GPa up to 7 GPa, combined with a low density – up to 2.55 Mg/m^3, exceptionally 3–4 Mg/m^3 for some of the ceramic fibres. Matrices usually have a much lower modulus and strength, 2–4 GPa and up to 100 MPa respectively with a density in the range 1–1.5 Mg/m^3. Thus specific properties – i.e. the ratio of the property to the density or specific gravity – may be very large, easily exceeding those of the metals. Fibre types are being actively developed particularly, for oriented polymers, carbons and ceramics, and the upper modulus and strength values will no doubt steadily increase.

It is possible to control the properties of a composite in several ways; by the selection of the fibre and matrix, the degree of bonding between the two, the length of the fibre, the relative proportions of fibre and matrix and the concentration of the fibres.

Fabrication involves impregnating the fibres with the matrix, consolidating the two and transforming the matrix to a continuous solid of the required shape by cross-linking (for a thermoset) or melting and resolidification (for a thermoplastic). Thermosets, depending on their type, react from room temperature upwards and may require a post cure at a temperature related to the heat distortion temperature (HDT) of the resin, i.e. up to 300°C. Thermoplastics require processing at a temperature which may approach 400°C. Processing times vary from a matter of seconds for stamping a reinforced thermoplastic up to tens of hours for the more complex, high temperature performance, thermosets. The pressures required for consolidation range from atmospheric to a few tens of MPa. Thus it is possible to fabricate excellent composites by hand with the aid of simple equipment. Alternatively sophisticated robotic equipment or computer controlled filament winding or tape laying machines may be used.

Composites based on glass fibre, known as GRP, have been available for five decades. Estimated consumption in 1990 was very approximately 1,000,000 tonnes each in Western Europe and USA and 450,000 tonnes in Japan. Much less composite material based on organic, carbon and ceramic fibres is used, approximately 12,000 tonnes worldwide in 1990. However, the properties of these fibres are such that composites produced from them, known as advanced composites, are extensively used in aerospace, high performance engineering and sports goods. Consequently, though the volume market may be small, the value of the products is very substantial.

3.2 THE ROLE OF THE FIBRE AND MATRIX

To gain some insight into the nature of a composite material consider a loose bundle or tow of carbon fibres. This will typically contain from 1,000 to 12,000 aligned continuous fibres each with a diameter of approximately 10 μm. A similar description can be given in terms of glass or aramid fibres though the diameter of the fibre and structure of the bundle will differ somewhat. Note that glass fibres are often described in terms of their tex, which is the weight in grams of 1000 m of filament or yarn. In some ways the tow of fibres resembles a section of rope, though fibres in the latter are twisted together. As the individual carbon (or glass or aramid) fibres have a high tensile strength and modulus the tensile properties of the tow are excellent but the structure has essentially no resistance to bending, shear, or torsion, or compressive loads. In each of these modes of stressing individual fibres, apart from some minimal frictional interaction, behave individually. If the bundle of fibres is saturated with a continuous solid polymer, the bending, shear and compressive performance of the material will be greatly enhanced. A detailed analysis shows that the stress in a fibre builds up to a maximum, peak, or constant plateau value which depends on stress transfer between the two components and the aspect ratio (ratio of fibre length to diameter) of the fibre. Typically the aspect ratio must exceed 10:1 to 1000:1 or more for maximum fibre stressing, and hence the efficient utilization of the fibre, to be attained.

Apart from promoting stress transfer and thus contributing to many composite properties, the matrix keeps the fibres oriented in the required directions, protects the fibre surface from damage, moisture ingress, and other environmental effects and, by modifying the progression of cracks through the material, may increase the overall work of fracture considerably.

■ 3.3 FIBRES

■ **Mechanical properties**

3.3 FIBRE PROPERTIES

The problems of making measurements on 10 μm diameter fibres are considerable, one difficulty with brittle fibres being a selection effect due to breaking weaker fibres when mounting them. Work has essentially been confined to the determination of fibre density, modulus, strength and the stress/strain curve.

In certain cases, tensile properties are measured using a thin strand of highly aligned, carefully spaced fibres in an epoxy matrix and the values of the fibre modulus and strength determined by scaling the results up to 100 v/o of fibre. It is arguable that this is a more realistic way of obtaining properties than single fibre measurements, since in practice fibres are used in a composite.

Some fibres, e.g. organic and carbon ones, are anisotropic in their behaviour. The transverse, compressive and shear properties of such materials, as well as thermal and electrical properties, have to be estimated by measuring the appropriate parameter for a well-made composite and then using a model which fits the data to obtain the appropriate fibre properties.

Although most types of fibre do not exhibit fully plastic behaviour, the stress/strain characteristic of carbon and organic fibres is not linear to failure. The modulus of carbon fibres may increase with increasing extension and hence ideally one should distinguish between tangent and secant modulus (section 1.4).

Some fibre properties, assembled from manufacturers' trade data are listed in Table 3.1. Carbon fibres have been divided up, somewhat arbitrarily, following Lovell (1991). This is necessary because of the number of different fibre types available. Frequently the data sheets do not distinguish between tangent and secant modulus, though for Kevlar the modulus values are 1% secant ones. Properties do vary between sources and all those in Table 3.1 are indicative rather than absolute.

Maximum temperature

The oriented polyethylene fibres have a maximum working temperature of approximately 90°C and aramid fibres of 180°C, though the decomposition temperature of the latter is > 500°C. The tensile strength of glass fibres begins to decrease between 220 and 260°C, falling to 50% of its room temperature value by 480–560°C depending on the type of glass. Carbon fibres will oxidize in air at temperatures between 300 and 400°C. The other fibres listed have considerably higher upper working temperatures ~1000–1400°C.

Table 3.1 Mechanical properties of fibres

fibre type	density (Mg/m^3)	tensile modulus (GPa)	tensile strength (GPa)	specific modulus (GPa)	specific strength (GPa)
glass					
E-glass (Vetrotex)	2.52–2.62	73	3.4	28	1.31
R-glass (OCF)	2.55	86	4.4	33.7	1.73
S-2 glass (OCF)	2.49	86	4.5	34.5	1.81
aramid					
Kevlar 129 (du Pont)	1.44	75	3.32	52	2.31
Kevlar H_p	1.44	100	2.96	69.4	2.06
Kevlar 49	1.45	117	2.76	80.7	1.9
Kevlar H_m	1.47	160	2.4	108.8	1.63
Twaron (Akzo)	1.44	80	2.8	55.6	1.94
Twaron HM	1.45	115	2.8	79.3	1.93
Technora (Teijin)	1.39	73	3.4	52.5	2.45
oriented polyethylene fibre					
Dyneema (Dyneema VOF)	0.97	87	2.7	90	2.78
Spectra 900 (Allied Signal)	0.97	117	2.65	121	2.73
Spectra 1000	0.97	172	3.09	177	3.19
carbon					
high strength/strain	1.7–1.9	160–250	1.4–4.93	84–137	0.74–2.9
intermediate modulus	1.7–1.83	276–317	2.34–7.07	150–186	1.28–4.15
high modulus	1.75–2.0	338–436	1.9–5.52	169–249	0.95–3.15
ultra high modulus	1.87–2.0	440–827	1.86–3.45	220–442	0.93–1.84
alumina					
Safimax SD (ICI)	3.3	300	2	90.9	0.61
Safimax LD	2	200	2	100	1
Saphikon (Saphikon)	3.97	462	4	116.4	1
Altex (Sumitomo Chemical Engineering Co)	3.25	210	1.8	64.6	0.55
alumina/boria/silica fibre					
Nextel 440 (3M)	3	186	2	62	0.67
Nextel 480	3.05	220	2	72	0.66
Nextel 312	2.7	152	1.72	56.3	0.64
silicon carbide					
Sigma (BP Metal Comp.)	3.4	390	3.5–3.9	114.7	1.03–1.15
Nicalon (Nippon Carbon)	2.55	193	2.7	75.7	1.06
Tyranno (Ube Industries)	3	400	3.45	133.3	1.15

■ 3.3 FIBRES

■ **Mechanical properties**

Properties of fibres

A useful way of illustrating the range of fibre properties is to plot breaking strength against modulus (Figure 3.1). The same means of determining strength and modulus may not have been used in each case. The two higher results for glass are for virgin filaments while the aramid figures are for twisted tow. The results for carbon fibres are just some of those available and have been chosen to illustrate the range of properties of available fibres. Lines of constant strain have been added.

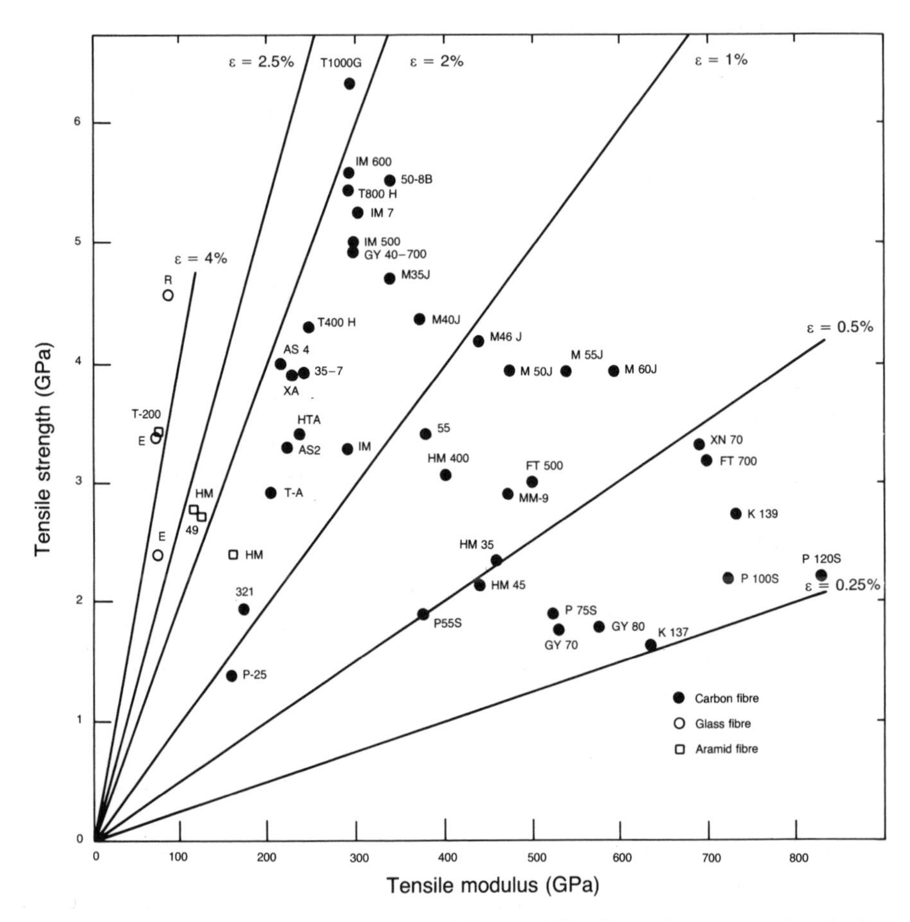

Figure 3.1 Fibre strength as a function of the modulus for carbon, aramid and glass fibre.

Suppliers

Carbon fibres

Akzo Fibres and Polymers Division Products	Amoco Performance Inc.
HTA	P 120S
HM 35	P 100S

3.3 FIBRES

Mechanical properties

HM 45
IM 600

P 75S
P 55S
P 25

BASF Structural Materials

GY 70
GY 80
GY 40-700

BP Chemicals Hitco

35-7
50-8B

Courtaulds

HM 400
XA
IM

Fiber Materials Inc.

55

Hercules Advanced Materials and Systems Company

AS2
AS4
IM7

Mitsubishi Kasei Corporation

K139
K137
T-A
MM-9
321

Nippon Petroleum Co. Ltd.

XN 70

Toho Rayon Co. Ltd.

IM400
IM500

Tonen Corporation

FT700
FT500

Toray Industries Inc.

M60J
M55J
M50J
M46J
M40J
M35J
T1000
T1000G
T800H
T400H
T300

Glass fibres

(Several sources)

E

Vetrotex

R

Aramid fibres

Akzo Fibres and Polymers Division

HM

EI duPont de Nemours and Co. Inc.

Kevlar 49
Kevlar H_m

Teijin Ltd.

T-200

Note: Courtaulds no longer manufactures carbon fibre. There are similar products available from other manufacturers.

■ 3.3 FIBRES

■ Mechanical properties

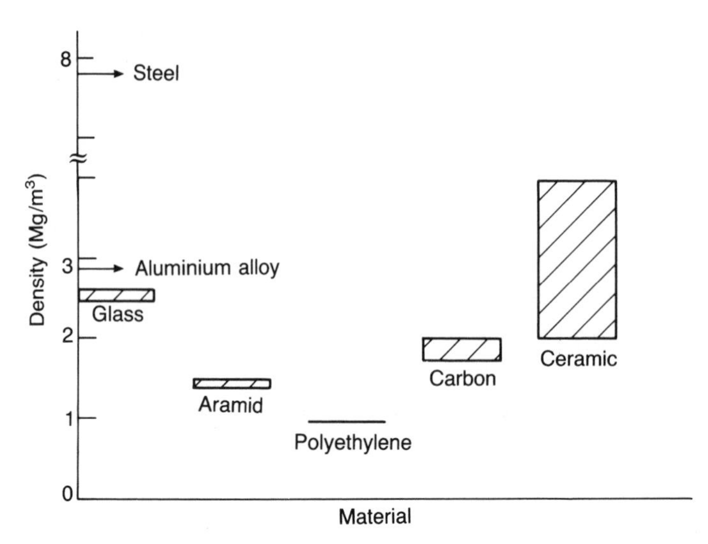

Figure 3.2 Density for glass, aramid, polyethylene, carbon and ceramic fibres compared with aluminium alloy and steel.

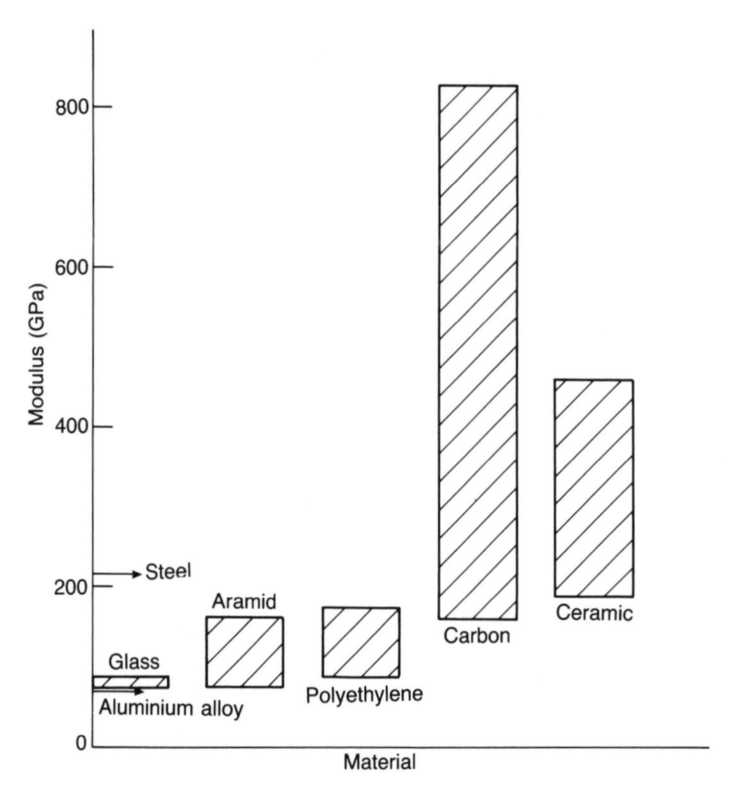

Figure 3.3 Modulus for glass, aramid, polyethylene, carbon and ceramic fibres.

3.3 FIBRES

Mechanical properties

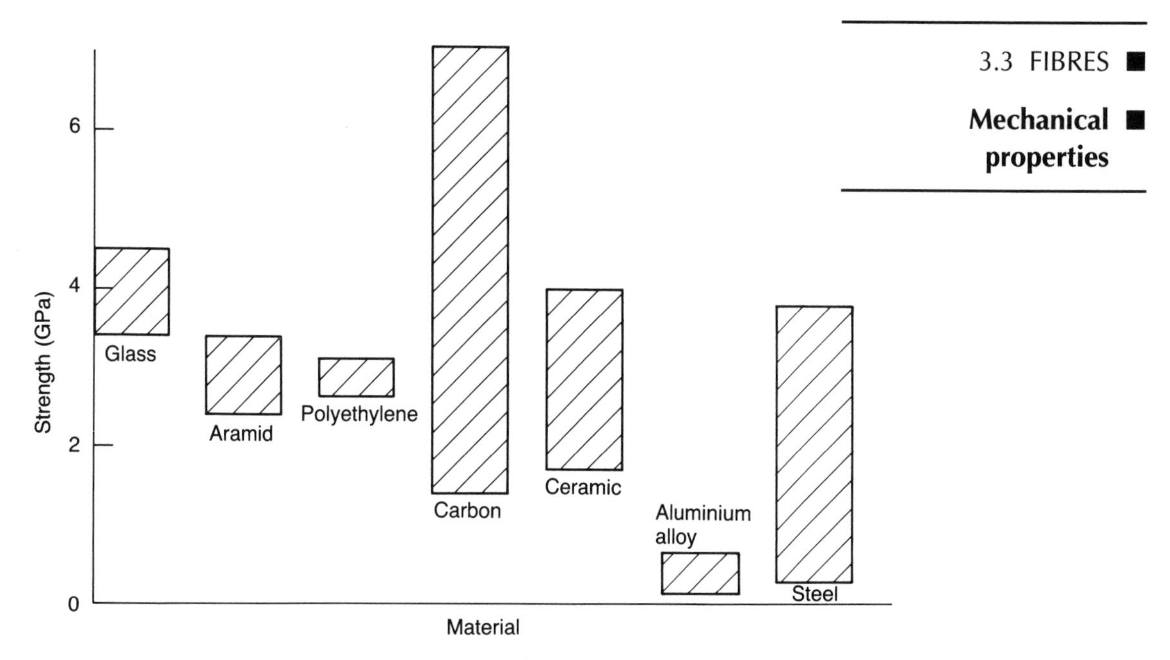

Figure 3.4 Strength for glass, aramid, polyethylene, carbon and ceramic fibres.

The information in Table 3.1 is shown in graphical form in Figures 3.2 to 3.4. These figures give an indication of the relative densities, moduli and strengths of the reinforcement. Values for aluminium alloy and steel are added for comparison. The strengths for glass refer to virgin filaments. The wide range in carbon and aramid fibre moduli and strengths is clear and to emphasize this the data for carbon and aramid fibres in Table 3.1 are replotted in Figures 3.5 and 3.6. The tendency for strength and strain to failure to decrease as modulus increases should be noted.

Source

Trade data.

■ 3.3 FIBRES

■ **Mechanical properties**

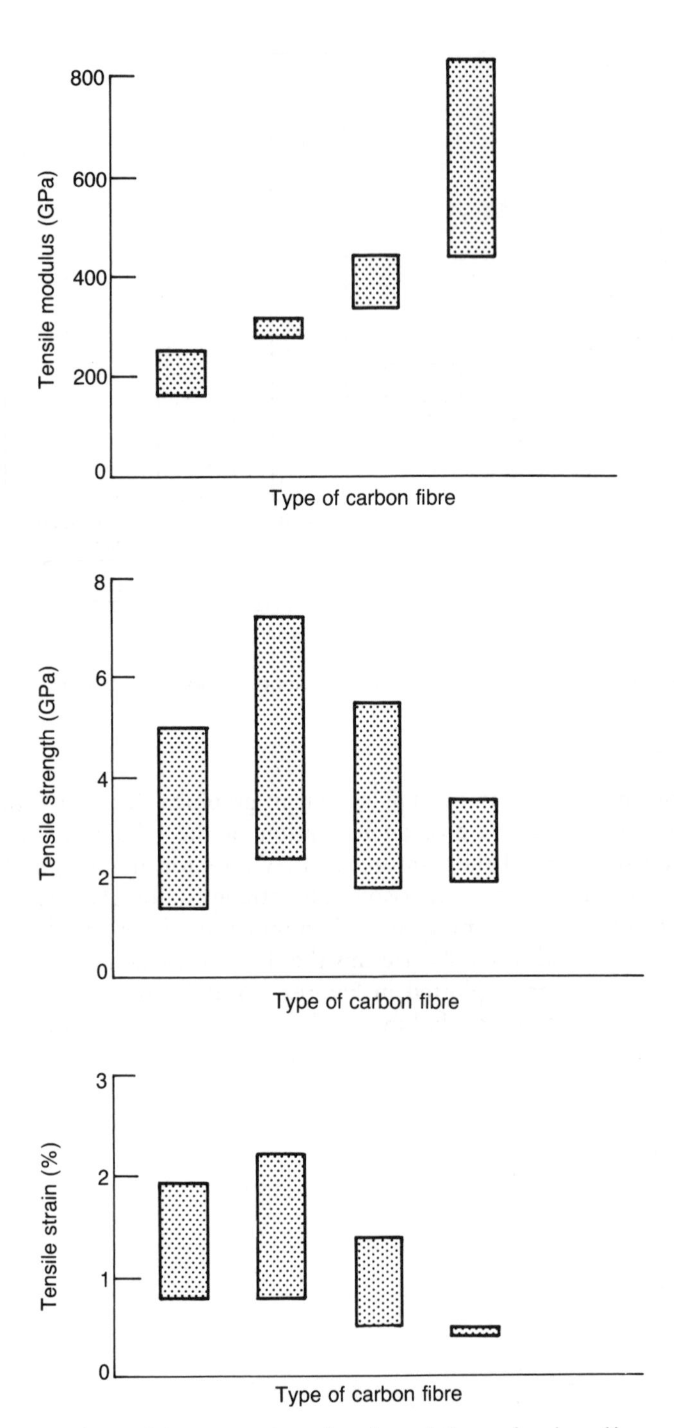

Figure 3.5 Tensile modulus, strength and strain-to-failure of carbon fibres.

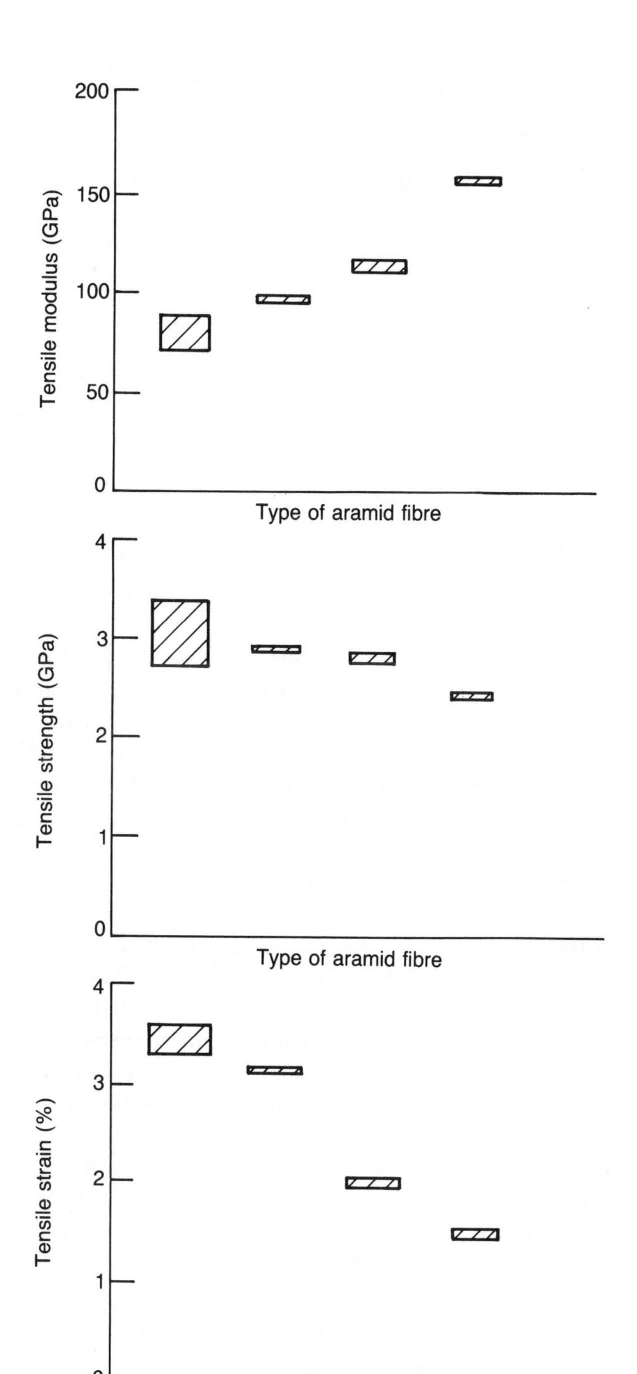

Figure 3.6 Tensile modulus, strength and strain-to-failure of aramid fibres. Note the narrower range of fibre properties.

■ 3.4 FIBRE FORMS

■ **Rovings, mats and weaves**

3.4 PHYSICAL FORM OF THE REINFORCEMENT

The handling of single fibres, thinner than a human hair and many hundreds of metres in length, to produce, economically, a reinforced structure would be virtually impossible. Consequently once the fibres have been surface-treated to enhance bonding and any protective coatings applied they are combined into tows consisting, for carbon fibres, of up to 12,000 individual filaments. Other terms, more commonly used for glass fibres, are 'strand' which is a compact bundle of fibres, 'roving' which refers to an untwisted group of strands, or 'yarn' which is a group of fibres held together by twisting. The fibres are often sold as tows or rovings which are wound onto a reel. This type of material can be used to fabricate bars and plates but it is essential to keep the fibre aligned when doing so. A much better way of using tows or rovings is in filament winding, in which a tow or a number of tows, impregnated with a thermosetting resin, are wound onto a shaped mandrel to produce articles with axial symmetry.

All types of fibre or combinations of fibre in the form of tows, rovings or yarns can be woven with a very wide range of patterns with the fibres essentially aligned in two directions. The yarns are crimped where they interface and this causes the strength and modulus of the fabric to be lower than would be expected from the properties of the fibres. Mats consist either of chopped rovings – typically 25–30 mm in length – or a continuous length of yarn laid in a plane in a swirl pattern (Figure 3.7). In both cases an organic binder holds the material together. Figure 3.8 shows diagramatically examples of a woven pseudo unidirectional, bidirectional, and a woven, needled, product. Further details on all types of mat and fabrics are given by Mayer (1993).

The advantages of fabrics are in the handling of the material when fabricating a composite so that fibre alignment is maintained and fibres are draped to follow the shape of a mould or tool, improved off-axis properties (essentially isotropic for a mat) and possibly better through-thickness properties.

Another type of product is a preform in which the fibres, whether in the form of a mat, fabric or some other construction, are initially formed to a, possibly, complex shape and stabilized with a binder. The preform is placed in a mould and a thermosetting resin injected, sometimes with vacuum assistance, to produce a component (section 11.3). The advantage of this approach is the ability to maintain the fibres in position during a complex moulding operation, increased product quality and improved productivity.

An alternative approach to building up components in which the fibres are aligned in various directions, which is often favoured with carbon fibres, is to produce a pre-impregnated material, known as a prepreg. Unidirectional fibres are spread out thinly and impregnated with a thermosetting resin which is partially cured or B-staged. The sheet or roll material (which is a fraction of a millimetre in thickness)

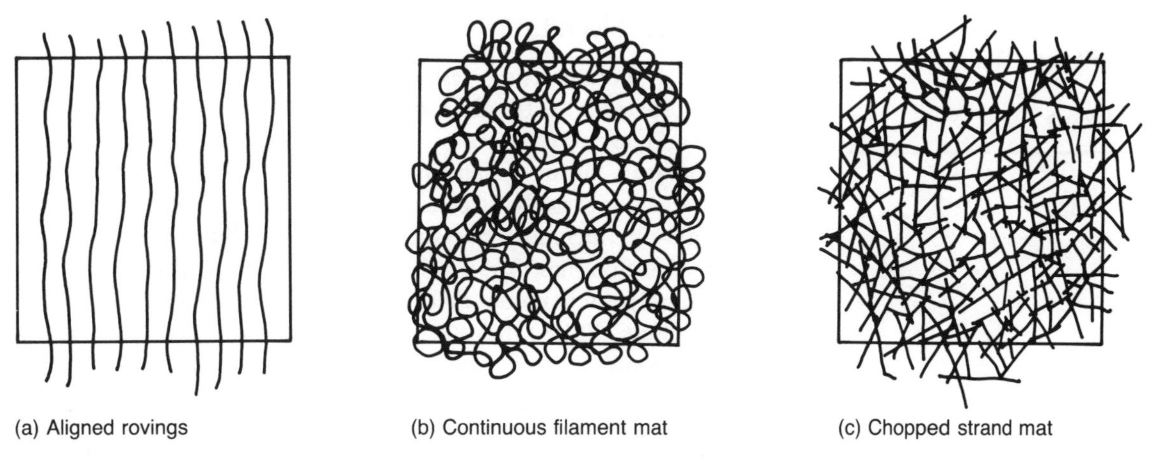

Figure 3.7 Examples of mat fabrics and a unidirectional material. (Reproduced with permission from Quinn, J.A. (1988), *Design Manual of Engineered Composite Profiles*, Fibreforce.)

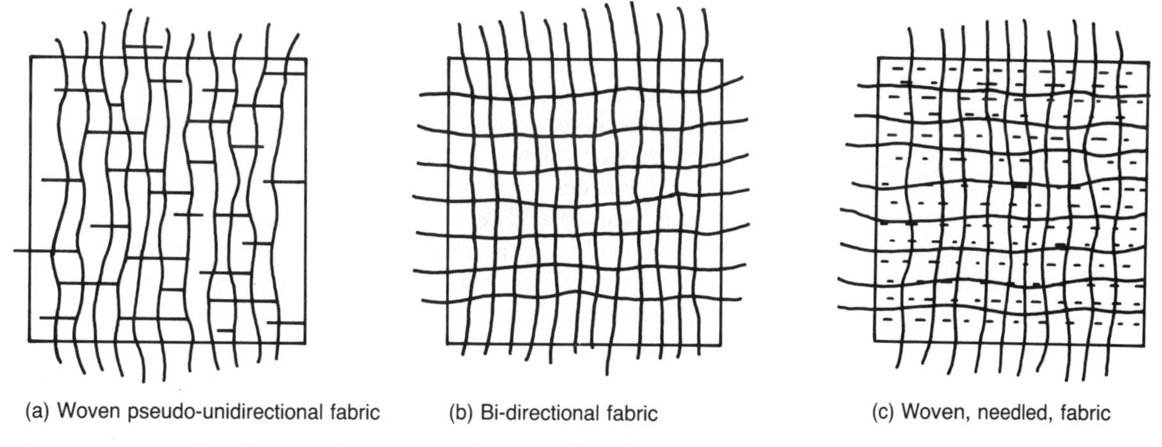

Figure 3.8 Examples of woven fabrics. (Reproduced with permission from Quinn, J.A. (1988), *Design Manual of Engineered Composite Profiles*, Fibreforce.)

and is described by the fibre or resin content, fibre type, etc., must be kept at a sub-zero temperature to stop the resin cure progressing. Laminates with calculable in-plane thermo-mechanical properties are made by cutting and stacking the material. The prepreg method can be used with thermoplastics and woven fabrics or to make tapes for use in winding.

The fibre volume loading, V_f, is usually between 55 and 65 v/o for an aligned fibre composite or laminate. The value may be lower for moulding compounds and GRP and in these cases it is often quoted as a weight percentage.

■ 3.5 MATRIX PROPERTIES

■ **Resin**

3.5 MATRIX PROPERTIES

The principal thermosetting resins used are phenolics, vinyl esters, polyesters (especially with glass fibres), and epoxies, bismaleimides, polyimides and polystyryl pyridine – all used with carbon fibres. Low or lower temperature curing systems must be employed with organic fibres to avoid damaging these by excessive heating during fabrication.

Thermoplastics including polysulphones (PSU), polyether sulphones (PES), polyarylsulphones (PAS), polyphenylene sulphide (PPS), polyether imide (PEI), polyamide imide (PAI), polyether ketone (PEK), and polyether ether ketone (PEEK), have several advantages as matrices (see the glossary for common tradenames). These include a higher strain-to-failure and a higher work of fracture than thermosets, the ability to be shaped and formed rapidly and repeatedly if necessary, by the application of heat and pressure, the good higher temperature resistance of some aromatic based systems and none of the problems associated with the limited lifetime of thermosets, especially if these have been partly reacted or B-staged.

The main disadvantage of thermoplastics is that their viscosity at the melt temperature is approximately two orders of magnitude greater than that of a thermoset. If the temperature is raised to aid fibre impregnation the polymer will be degraded. To overcome this various techniques are used based on film stacking, the use of powerful solvents or powdered polymer material.

The choice between a thermoset and a thermoplastic is not always simple. Much work has been and is being carried out to toughen thermosets with rubber or thermoplastic additives; repeated heating and forming of a thermoplastic may result in a degradation of properties and the excellent fracture resistance of thermoplastics may not translate directly into good composite properties. At low, sub-zero, temperatures thermosets are tougher and thermoplastics more brittle and the former are to be preferred.

There are many varieties of thermoset, particularly polyesters and epoxies, available which give a range of properties and processing conditions. There are fewer high temperature thermosetting and thermoplastic systems. Many thermoplastics are compounded with short glass or carbon fibres or other fillers before sale.

Some typical properties of the two families of polymer are listed in Table 3.2. It must be most strongly emphasized that the information is again indicative and will depend upon the exact system used and the cure schedule. For further details the manufacturers' data sheets should be consulted. Among the important points are that all the systems have similar densities and relatively low strengths and moduli.

Ideally, thermosets should be post cured at or above the required temperature of use. This may not be practical for very large structures and then a prolonged period (i.e. up to one month or more) at 20°C will be necessary. The quoted temperature performance can only be attained if the resin is fully cured – see Table 9.7. The upper working temperature is

Table 3.2 Mechanical properties of resins

type	density (Mg/m^3)	modulus (GPa)	strength (MPa)	strain-to-failure (%)	continuous service temperature (°C)
thermosets					
polyurethane	1.2	0.7	30–40	400–450	80
phenolic	1.0–1.25	3.0–4.0	60–80	1.8	120–260
polyester	1.1–1.23	3.1–4.6	50–75	1.0–6.5	70–130
vinyl ester	1.12–1.13	3.1–3.3	70–81	3.0–8.0	80–150
methacrylate	1.08–1.1	2.4–3.0	60–90		80–120
epoxy	1.1–1.2	2.6–3.8	60–85	1.5–8.0	80–180
bismaleimide	1.2–1.32	3.2–5.0	48–110	1.5–3.3	180–200
polyimide	1.43–1.89	3.1–4.9	100–110	1.5–3.0	250–300
polystyryl pyridine	1.14	2.3–3.0	41–59	1.7–3.0	250–300
thermoplastics					
PP	0.9	1.1–1.6	31–42	100–600	110
PA	1.1	2.0	70–84	150–300	140
PSU	1.25	2.5	60–75	50–100	150
PES	1.37	3.2	84	40–80	180
PAS		2.9	83	40	160–200
PPS	1.36	3.3	84	4.0	100–200
PEI	1.27	3.0	105	60	170
PAI	1.4	3.7–4.8	93–147	12–17	230–260
PEK			110		130–250
PEEK	1.26–1.32	3.2	93	50	120–250

Notes: Modulus and strength properties are primarily tensile for thermoplastics. For more detailed information on thermosetting resins, see Chapter 4.

really a matter of end use and will depend on the applied stress, environment and mechanical performance expected from the material. The only satisfactory way of determining this temperature is by testing the complete component or structure under service conditions. Thus a system may have a higher temperature rating if used as a stress free coating *in vacuo* than if used stressed in a corrosive chemical environment. In addition, with polyesters and epoxies the upper working temperature depends very much on the particular system used. Some of the thermoplastics, PPS, PEK and PEEK, are semi-crystalline and their cooling after forming must be controlled to give the required degree of crystallinity and hence properties.

Unlike fibres it is relatively simple to make measurements on most matrices, the exceptions being some of the high temperature thermosets which are not suitable for bulk casting. Furthermore the materials are isotropic. There is some discussion as to whether a μm sized flake has the same properties as bulk material but little or no evidence to support the idea that the properties differ significantly.

Source

Trade data.

■ 3.6 COMPOSITES

■ **Anisotropy notation**

3.6 COMPOSITE PROPERTIES

It will be recalled from section 3.1 that a fibre composite is inherently anisotropic because of the disparity between the properties of the fibre and the matrix. For GRP the ratios of the longitudinal to the transverse modulus and strength are of the order of 4 and 28 respectively, while similar figures for an ultra high modulus carbon fibre composite are 56 and 45, respectively. Materials with such highly directional properties can be hard to handle initially, particularly when coupled with the inability of thermoset matrix systems to undergo plastic deformation.

A fairly complete list of composite properties, together with the

Table 3.3 Composite properties required

property	units	symbol
density	Mg/m^3	ρ
tensile		
longitudinal tensile modulus	GPa	E_{lt}
longitudinal tensile strength	MPa	σ_{lt}
strain-to-failure	%	ε_{lt}
transverse tensile modulus	GPa	E_{tt}
transverse tensile strength	MPa	σ_{tt}
strain-to-failure	%	ε_{tt}
major Poisson's ratio		ν_{lt}
minor Poisson's ratio		ν_{tl}
compressive		
longitudinal compressive modulus	GPa	E_{lc}
longitudinal compressive strength	MPa	σ_{lc}
strain-to-failure	%	ε_{lc}
transverse compressive modulus	GPa	E_{tc}
transverse compressive strength	MPa	σ_{tc}
strain-to-failure	%	ε_{tc}
major Poisson's ratio		ν_{ct}
minor Poisson's ratio		ν_{cl}
flexural		
longitudinal flexural modulus	GPa	E_{lf}
longitudinal flexural strength	MPa	σ_{lf}
strain-to-failure	%	ε_{lf}
transverse flexural modulus	GPa	E_{tf}
transverse flexural strength	MPa	σ_{tf}
strain-to-failure	%	ε_{tf}
interlaminar shear strength	MPa	ILSS
shear		
in-plane shear modulus	GPa	G_{lt}
in-plane shear strength	MPa	τ_{lt}
strain-to-non-linearity	radians	γ_{lt}
transverse shear modulus	GPa	G_{tt}
in-plane shear strength	MPa	γ_{tt}
strain-to-non-linearity	radians	τ_{tt}

Anisotropy notation ■

units and symbols, is given in Table 3.3. Certain of these are dominated by the fibre while others are much more dependent upon the matrix. Generally the longitudinal properties of continuous fibre composites, whether short- or long-term, are fibre dominated. Examples are modulus, strength, strain-to-failure, thermal expansion and thermal conductivity. Given the appropriate properties and proportions of the fibre and matrix, the rule of mixtures can be used to estimate the unidirectional composite properties, (with the exception of the coefficient of expansion), with a reasonable degree of accuracy. The equation is of the form

$$X_c = X_f V_f + X_m V_m$$

Table 3.3 *Continued*

property	units	symbol
fracture		
longitudinal Izod impact strength (unnotched)	J/m	
longitudinal Izod impact strength (notched)	J/m	
transverse Izod impact strength (unnotched)	J/m	
transverse Izod impact strength (notched)	J/m	
longitudinal Charpy impact strength (notched)	J/m^2	
transverse Charpy impact strength (notched)	J/m^2	
drop weight impact strength	J	
energy absorption	various	
critical work of fracture (I)	J/m^2	G_{IC}
critical work of fracture (II)	J/m^2	G_{IIC}
critical work of fracture (III)	J/m^2	G_{IIIC}
critical stress intensity factor (I)	$MN/m^{3/2}$	K_{IC}
critical stress intensity factor (II)	$MN/m^{3/2}$	K_{IIC}
critical stress intensity factor (III)	$MN/m^{3/2}$	K_{IIIC}
thermal		
longitudinal CTE	$°C^{-1}$	α_l
transverse CTE	$°C^{-1}$	α_t
longitudinal CTC	W/m.K	λ_l
transverse CTC	W/m.K	λ_t
specific heat	J/kg.K	C_p
maximum working temperature	°C	
absorption coefficient		
emission coefficient		
electrical		
longitudinal electrical conductivity	S/m	σ_{el}
transverse electrical conductivity	S/m	σ_{et}
longitudinal electrical resistivity	Ω.m	ρ_{el}
transverse electrical resistivity	Ω.m	ρ_{et}
dielectric constant		ε'
tan δ		

■ 3.6 COMPOSITES

■ **Anisotropy notation**

where c, f, m refer to composite, fibre and matrix respectively. Usually the second term on the rhs can be ignored.

The longitudinal compressive modulus and strength, although markedly affected by the properties of the fibre, can also be strongly influenced by the type and behaviour of the matrix because of the lateral support that the latter gives to the fibre.

The transverse and shear properties of a unidirectional composite are principally dependent upon the behaviour of the matrix and fibre interface. Consequently they can be orders of magnitude less than their longitudinal counterparts. For GRP the shear modulus and strength are typically 5 GPa and 60 MPa and for ultra high modulus carbon fibre composite 4 GPa and 50 MPa, respectively.

Fracture properties are complex, depending on the maximum strain energy that can be stored prior to failure, delamination or debonding and whether the material is or is not described by linear elastic fracture mechanics.

The service temperature must be below that at which the material begins to soften, oxidize or otherwise degrade. With the exception of the organic fibre reinforcements the limiting phase is usually the matrix (Table 3.2). In addition chemical resistance is determined by the nature of the matrix since this is the component that first experiences the environment.

3.7 DESIGN AND MANUFACTURE

In designing with a composite material it is essential to identify the directions and magnitudes of the stresses accurately and place the fibres as required. For thermosetting matrix composites the maximum stress should not, even locally, exceed the elastic limit.

The fibre/resin bond strength is critical. If it is too weak, the resulting composite has poor mechanical properties and may be subject to excessive damage by moisture, though in an impact situation it will be able to absorb a lot of energy because of delamination and debonding. On the other hand if the bond is too strong, the material will behave in a brittle manner. To get an adequate bond strength requires matching the chemistry of the fibre surface to that of the matrix. To do this it is best to approach the fibre manufacturer with details of the matrix to get the correct fibre sizing formulation as many proprietary ones are available.

Certain materials, for example glass fibre mats and moulding compounds, may be formed directly by moulding or stamping, but in many applications the final product is made in one operation starting from the basic fibre and matrix.

Problems can arise in either case due to flow forces in the resin reducing the effective length of short or intermediate fibres and causing preferential alignment or misalignment of the fibres. These effects can be reduced by careful mould design and process monitoring and by understanding the flow of resin whether this is being injected into a mould or is present already as a partially reacted or thermoplastic material.

The matrices, both thermosetting and thermoplastic, have processing or curing schedules designed to build up the proper molecular structure, degree of crystallinity, cross-linking and chemical reactions and to aid with removal of solvents, etc., in the solid. If the sequence of temperatures and pressures is not adhered to or is incorrect, an inferior product will result. In some cases when using advanced thermosetting resins it may be necessary for the fabricator to optimize the process themselves.

Other sources of difficulties that can arise when manufacturing composites are due to moisture absorption (aramid fibres and certain hardeners are prone to this and the former should be dried immediately prior to use), contamination of mould or prepreg surfaces by oil or grease (including contamination from human skin), and using either thermoset resins, prepregs or sized fibres which are too old. Resins may deteriorate and pre-reacted resins cross-link to such an extent that they can no longer be formed by pressure and heat. The size on glass fibre can age and this may result in a higher than expected void content in a composite.

Generally if the performance of a composite or component appears substandard the processing route should always be carefully examined. Small changes in this can lead to significant increases or decreases in the quality and performance of an artefact.

3.8 HEALTH AND SAFETY

Reputable manufacturers and suppliers of resins, matrices and reinforcements provide extensive advice on storing materials and health hazards that may be associated with their products. Further advice and information can be obtained from:

- British Plastics Federation, 6 Bath Place, Rivington Street, London, EC2A 3JE, UK, Tel. (44) 071 457 5000;
- COSHH Assessments, HMSO, London;
- Health and Safety Executive, Belgrave House, Greyfriars, Northampton, UK, Tel. (44) 0604 21233.

3.9 REFERENCE INFORMATION

Lovell, D.R. (1991) *Carbon and High Performance Fibres Directory*, 5th Edition, Chapman & Hall, London.

General references

Ashton, J.E. *et al.* (1969) *Primer on Composite Materials: Analysis*, Technomic Publishing Company Inc., Lancaster, USA.

Bowen, D.H. (1986a) Manufacturing methods for composites. *Metals and Materials*, September 1986, 584–8.

Bowen, D.H. (1986b) Applications of polymer matrix composites. *Metals and Materials*, December 1986, 776–9.

Davidson, R. (1986) Performance characteristics of composite materials. *Metals and Materials*, October 1986, 651–5.

Hancox, N.L. (1986a) Principles of fibre reinforced composites. *Metals and Materials*, May 1986, 285–7.

Hancox, N.L. (1986b) Matrices for composite materials. *Metals and Materials*, July 1986, 435–7.

Hughes, J.D.H. (1986) Fibres for reinforcement. *Metals and Materials*, June 1986, 365–8.

Hull, D. (1981) *An Introduction to Composite Materials*, Cambridge University Press, Cambridge.

Jones, R.M. (1975) *Mechanics of Composite Materials*, McGraw-Hill/Kogakusha Ltd, Tokyo.

Mayer, R.M. (ed.) (1993) *Design with Reinforced Plastics*, Design Council, London.

Concise Encyclopaedia of Composite Materials (1989) Pergamon Press, Oxford.

Handbook of Composites (1983) Volume 4, Chapter 1, North Holland, Amsterdam.

The Ashton *et al.* and Jones references are concerned with the theory of laminates. The two Bowen, two Hancox, and Davidson and Hughes articles are short, popular articles about different aspects of composite materials.

4 MECHANICAL PROPERTIES OF RANDOM AND FABRIC-REINFORCED RESINS

SUMMARY

The short-term mechanical properties of resins reinforced with mats or fabrics are discussed. Grouping is by resin type for the principal types used for structural applications. In contrast to properties in the fibre directions there are little data for transverse and out-of-plane directions.

4.1 INTRODUCTION

Reinforcement

Mats and fabrics are extensively used as these are easy to lay-up and to wet-out and so form the basis of the GRP industry.

Combinations of these two types of reinforcement exist in the form of 'combination mats' and their properties tend to be the aggregate of the individual materials. Since many permutations are possible and few preferred types have emerged within the industry, some characteristic data are described.

Random roving composites are produced by the spray-up method (section 11.2) in which an intimate mixture of roving and resin is sprayed into, or onto, a mould. Properties of aligned rovings are considered in the next chapter.

Property level

For some thermosetting resins, notably polyesters, a variety of catalysts and accelerators is available and it is implicitly assumed that properties are always characteristic of fully cured composites (section 12.3).

Unless otherwise indicated, the data are given for the principal fibre direction(s) of the composite.

In general, it appears from the data presented here that the tensile modulus is higher than the flexural modulus, whereas flexural strength is higher than tensile strength (*cf.* for example Figures 4.6 and 11.2). This is explicable in terms of the roles that the resin and reinforcement play in various types of loading.

4.2 POLYESTER RESINS

Cast resins

Typical properties of the three principal types are listed in Table 4.1.

Resins
The resins are:

- orthophthalic;
- isophthalic;
- isophthalic neo pentyl glycol;

cured in accordance with manufacturers' recommendations.

Observations
These three groups form the basis of most resins. Many blends and variations are available depending upon, *inter alia*, the nature of the application and the process used.

A major difference between the groups is their elongation at break, particularly as a high elongation aids toughness and impact properties (section 6.2).

Resins with similar heat distortion temperatures are shown in Table 4.1. However, higher temperature versions of these resins are also available, for example, iso-polyesters have the highest HDT (*c*.130°C), iso-NPG, somewhat lower (*c*.110°C) and ortho-polyesters the lowest (*c*.100°C).

Post curing at or above the required temperature of use may be necessary, and indeed required, depending upon the type of application (Table 9.7).

Table 4.1 Typical properties of cast polyester resins

		ortho-	iso-	iso-NPG-
viscosity	mPa.s	350–400	280–350	270–300
density	Mg/m^3	1.20–1.23	1.19–1.20	1.14–1.16
tensile strength	MPa	50–70	70–75	70–73
tensile modulus	GPa	3.8–4.6	3.5–3.6	3.1–3.4
elongation at break	%	1.6–2.3	3.0–5.0	3.0–6.5
flexural strength	MPa	90	130	135
flexural modulus	GPa	4.0	3.7	3.0
Barcol hardness		45	45	35–45
heat distortion temperature	°C	62–72	75–78	76–90
in-mould shrinkage	%	5.5–6.5	7–8	8
water absorption (28 days)	%	0.8	0.8	0.5

Notes: Water absorption is measured according to ISO 62 using a plaque size 50 mm × 50 mm × 3 mm. If any other size is used, the volume pick-up needs to be converted to mass pick-up using the formula $m = \rho V$. Varying exposure times could also lead to different values.

■ 4.2 POLYESTER RESINS

■ **Cast resins**

Properties

Addition of urethane acrylate

This polymer is fully miscible in polyester resin over the entire range of composition from 100% polyester to 100% urethane acrylate. The effect on modulus, strength and failure strain is illustrated in Figure 4.1.

There is a clear trade-off between modulus and strength on one hand and failure strain on the other. The consequence is that the more flexible the resin, the lower the strength and modulus.

The effect on impact properties is described in section 6.2 and Figure 6.2.

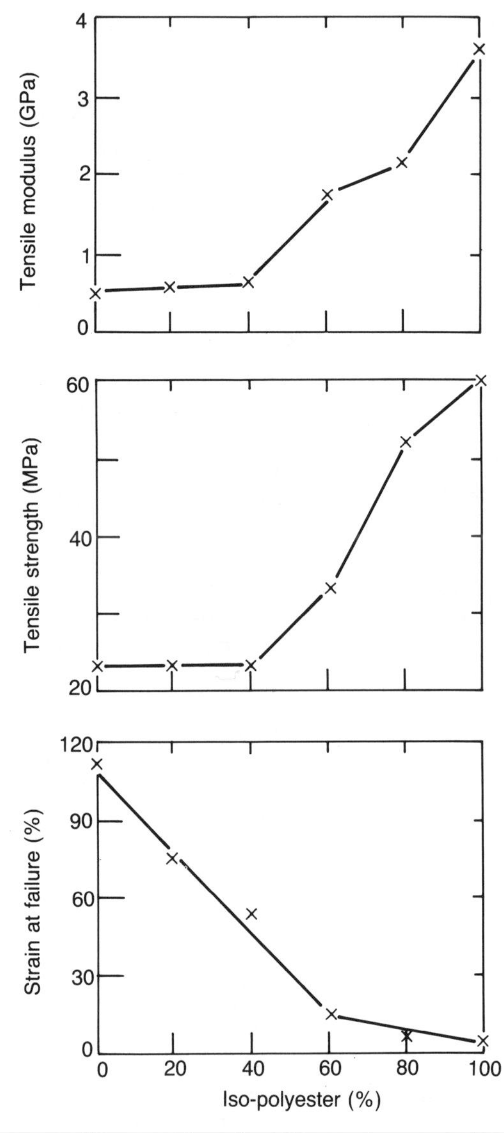

Figure 4.1 Effect on tensile properties of adding urethane acrylate to polyester resin (Norwood).

Design implications

Polyesters are very widely used because of their wide property range and ease of handling and processing. Shrinkage on moulding can be a problem and reinforcing with fibres (and/or fillers) will reduce the shrinkage values of the cast resins appreciably.

Allowance can generally be made for shrinkage of composites so that components can be moulded direct to the final size.

It is possible to mould very large components because of the availability of delayed action catalysts.

Sources

R. Arvesen at Jotun Polymer.
L. Norwood at Scott-Bader.

■ 4.2 POLYESTER RESINS

■ **Glass size and tex**

Spray mouldings

Effect of glass size and tex

The effect of glass size and tex on spray-up moulding is shown in Figures 4.2 and 4.3.

Materials

The *materials* are:

- glass – chopped rovings from various manufacturers having different sizes or tex;
- resin – ortho-polyester, Norpol 44M (Jotun Polymer).

Manufacture

Manufacture is achieved by spray moulding.

Observations

The effect of glass size causes a variation in property level, the extent of which will depend upon the type of size, process and property being measured. Tex is also important for spray mouldings with finer tex (40) giving better properties than a coarse tex (80) for a given level of reinforcement.

Design implications

Care should be taken when selecting a glass/resin combination with which the designer or moulder is unfamiliar; property levels should then be checked on test coupons. If a required value has to be attained

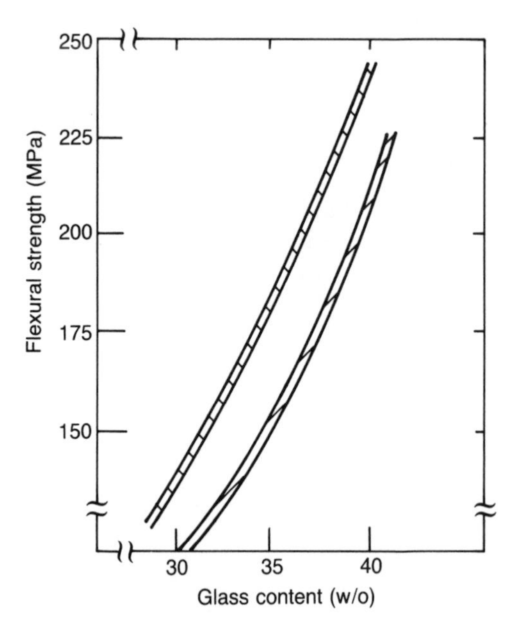

Figure 4.2 Effect of two different glass sizes on flexural strength of laminates made from spray rovings (Arvesen).

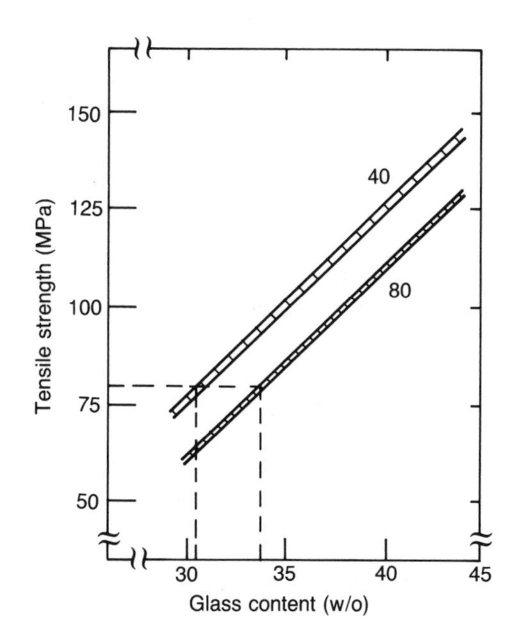

Figure 4.3 Effect of tex on tensile strength of laminates made from spray rovings (Arvesen).

to meet a particular standard, then material selection becomes important.

Source

R. Arvesen at Jotun Polymer.

■ 4.2 POLYESTER RESINS

■ **Spray mouldings**

Flexural properties

Flexural properties of spray mouldings

The flexural strength and modulus obtainable from spray mouldings is shown in Figure 4.4.

Materials

The *materials* are:

- glass – chopped rovings;
- resin – ortho-polyester, Norpol 44M (Jotun Polymer).

Manufacture

Manufacture is achieved by spray moulding.

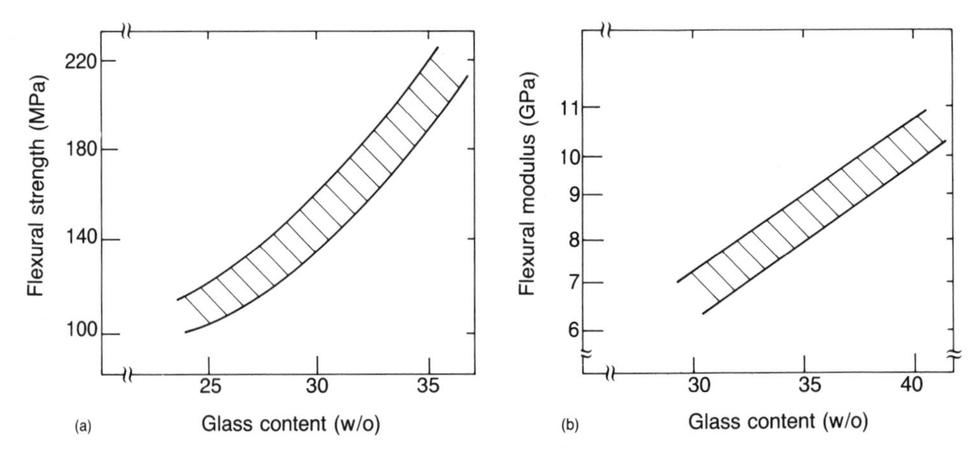

Figure 4.4 Variation of (a) flexural strength, and (b) flexural modulus with glass content for spray moulding. Band covers spread in sizing and tex (Arvesen).

Observations

These graphs indicate the working range of glass content and flexural properties to be encountered in spray moulding. Other resins might give slightly different variations.

Design implications

There are limits to what can be achieved by spray moulding. The lower limit is set by the need to obtain adequate coverage throughout the moulding and the upper limit by the ability to achieve adequate compaction. For higher levels of properties, other types of reinforcement would need to be considered.

Source

R. Arvesen at Jotun Polymer.

Flexural properties of mats

Glass mat in the form of chopped rovings or continuous filaments can give a wide range of properties when processed by hand lay-up (Figure 4.5).

Materials

The *materials* are:

- glass – chopped strand mats (types A and B);
- resins – ortho-polyester, Norpol 44 (Jotun); iso-polyester, Norpol 72 (Jotun).

Manufacture

Manufacture was achieved by contact moulding by hand lay-up.

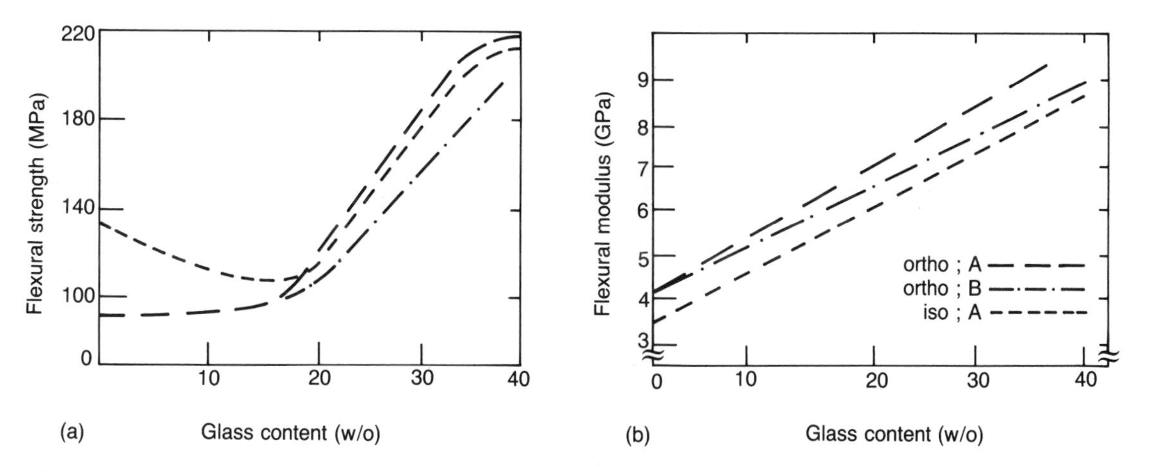

Figure 4.5 Variation of (a) flexural strength, and (b) flexural modulus with glass content for contact moulding (Arvesen).

Observations

Whilst the modulus increases linearly with fibre content, the composite strength may be reduced at low glass contents, particularly for iso-polyester resins. The variation in both resin type and mat type gives rise to a variation in properties similar to that of spray moulding.

Design implications

There are both lower and upper limits for effective reinforcement, particularly for strength. The reinforcement range is similar to that of spray rovings, but mats give a higher strength for a given glass content.

Source

R. Arvesen at Jotun Polymer.

■ 4.2 POLYESTER RESINS

■ **Chopped strand mats**

Mechanical properties by resin type

Property variation with chopped strand mats

There is a wide variety of chopped strand mats (CSM), which are either bonded together using an emulsion or a powder. The properties of a range of mats are listed in Table 4.2.

Materials

The *materials* are:

- reinforcement – 5 types of powder bonded CSM, 2 types of emulsion bonded CSM;
- resins – marine grade ortho-polyester (Scott-Bader), marine grade iso-polyester (Scott-Bader).

Manufacture

Manufacture was achieved by hand lay-up and contact moulding.

Observations

The mats give a higher tensile strength with iso-polyester than with ortho-polyester resin, and a correspondingly higher strain-to-failure.

Powder bonded mats tend to give a higher level of properties than emulsion bonded mats.

Design implications

A variety of mats is available. If mechanical properties are important then powder bonded mats are preferable to emulsion bonded mats and mats should be screened to obtain optimum properties.

Shear properties are listed in Table 4.4; the effect of adding urethane acrylate to the polyester matrix is listed in Table 4.10.

Table 4.2 Property variation of polyester resins with chopped strand mats

	mat	powder bonded		emulsion bonded	
	resin	ortho-	iso-	ortho-	iso-
glass content	w/o	29–30	29–31	29–30	29
tensile strength	MPa	90–118	90–130	76–92	107–110
tensile modulus	GPa	7.5–8.8	7.4–8.4	7.5–7.1	7.4–7.2
failure strain	%	1.5–2.0	3.7–4.2	1.2–1.7	2.1–4.1
flexural strength	MPa	165–243	177–230	169–173	192–199
flexural modulus	GPa	5.9–7.2	6.0–6.6	6.9–6.6	6.4–5.2
failure strain	%	3.1–3.6	3.0–3.5	2.4–2.6	3.0–3.8

Source

L. Norwood at Scott-Bader.

Properties of mats and fabrics

Typical properties of mats and fabrics across a range of glass contents are listed in Table 4.3

Materials

The *materials* are:

- glass – chopped strand mat (34w/o), combination mat balanced woven fabric plus random chopped layer (47–53w/o), combination mat plus unidirectional fabric (61w/o);
- resin – iso-NPG-polyester, high elongation, Norpol 20 (Jotun).

Manufacture

Manufacture was achieved by contact moulding through hand lay-up.

Observations

The data show how the modulus and strength increase with glass content and fibre orientation. This is in accordance with the predictions of the law of mixtures (section 1.3). The increase in glass content is achieved by the fabrics becoming increasingly orientated and finally biased in one direction (Figure 4.6); there will, however, be a corresponding reduction in properties in other directions.

Note that the high elongation at break of the matrix (Table 4.1) is not carried over into the composite.

The theoretical variation in tensile modulus, shear modulus and Poisson's ratio with glass content is shown in Figure 4.7 for comparison.

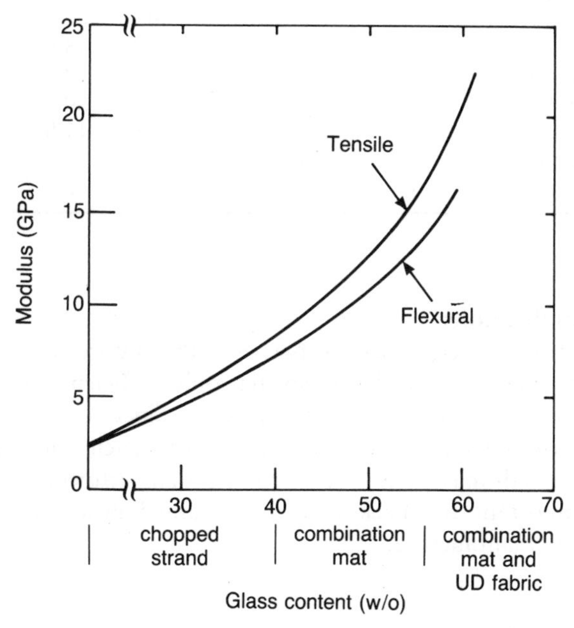

Figure 4.6 Effect of glass content on tensile modulus and flexural modulus for a variety of glass-reinforced laminates (Arvesen).

■ 4.2 POLYESTER RESINS

■ **Mats and fabrics**

Mechanical properties

Glass content

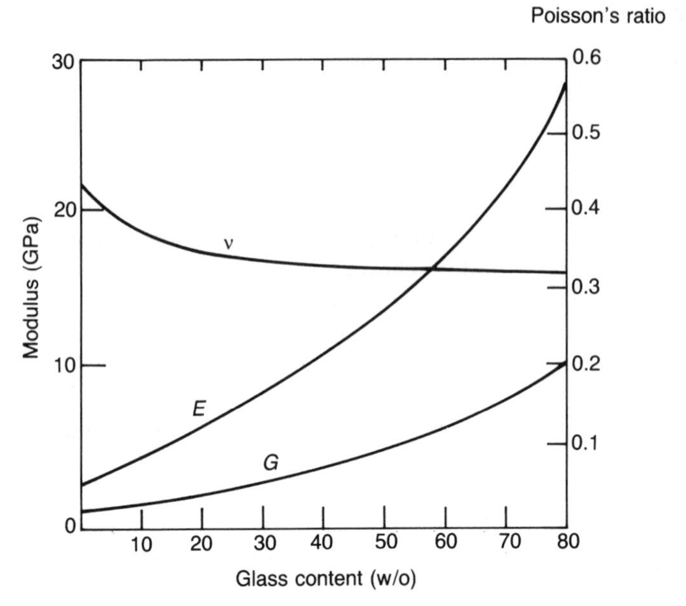

Figure 4.7 Theoretical variation in tensile modulus (*E*), shear modulus (*G*) and Poisson's ratio (ν) with glass content (Vetrotex).

Table 4.3 Typical properties of glass polyester laminates

glass content	w/o	0	34	47	53	61
tensile strength	MPa	73	100	200	240	420
tensile modulus	GPa	3.1	7.5	13.0	14.5	22.5
elongation at break	%	6.5	1.8	2.2	2.3	2.3
flexural strength	MPa	135	140	320	330	
flexural modulus	GPa	3.0	6.0	11.0	11.5	
tensile strength (out-of-plane)	MPa			15	15	16
interlaminar shear strength	MPa			35		

Design implications

The variation in mechanical properties with reinforcement content can be used to advantage in design so that only sufficient reinforcement is used to achieve a desired property level. Since GRP structures are generally stiffness limited compared with metals, achieving the necessary level of strength may be less of a problem than stiffness.

It is important to check that the assumed property level can be achieved by manufacture.

Sources

R. Arvesen at Jotun Polymer.
G. Creux at Vetrotex.

Shear properties

The properties in the non-fibre directions are very much lower than those along the fibre directions. Some data are listed in Table 4.4.

Materials
The *materials* are:

- glass – chopped strand mat, woven fabric, balanced warp and weft;
- resin – iso-polyester (Scott-Bader).

Manufacture
Manufacture was achieved by contact moulding and hand lay-up.

Observations
The properties away from the principal fibre direction(s) are much lower than along the fibre direction. Moreover, the measurements can be made in various ways as there are different types of shear loading.

Design implications
If shear or off-axis stresses are not negligible then the property levels need to be checked by an appropriate method. The design may need to be altered to reduce the stress levels or change the fibre orientation or alter the laminate stacking sequence.

Table 4.4 Measurements of shear properties

		powder bonded CSM	emulsion bonded CSM	woven fabric
glass content	w/o	32	31	57
in-plane (rail)				
shear strength	MPa	59 ± 13	56 ± 8	65 ± 2
shear modulus	GPa	2.4 ± 0.2	1.95 ± 0.3	1.78
(45° off-axis tension test)				
shear strength	MPa			56.4 ± 0.6
shear modulus	GPa			2.5 ± 0.1
out-of-plane (interlaminar)				
(short beam shear test)				
shear strength	MPa	26.4 ± 1.4	17.9 ± 0.8	32.4 ± 1.4
(offset 4-point bend test)				
shear strength	MPa	26.9 ± 2.7	19.7 ± 1.3	26 ± 1

Note: For test methods consult Mayer (1993).

Source
L. Norwood at Scott-Bader.

■ 4.2 POLYESTER RESINS

■ **High glass fraction**

Mechanical properties

Effect of high glass fraction

There is generally an upper limit (as well as a lower limit) to the positive influence of glass reinforcement. The lower limit has already been discussed for a random mat (Figure 4.5).

Materials

The *materials* are:

- reinforcement – unidirectional inlaid glass fabric, 97% warp, 3% weft, 960 g/m^2 (Flemings Laces);
- resin – iso-polyester (Scott-Bader).

Manufacture

Manufacture was achieved by contact moulding and hand lay-up.

Observations

The flexural modulus continues to increase above 50 v/o glass content although more slowly than the law of mixtures would predict. The strength, on the other hand, peaks at about 50 v/o and does not increase any further. The effect on the strain is such that this peaks at 50 v/o and then steadily decreases.

Design implications

Care must be taken in trying to achieve optimum properties above 50 v/o glass content if fabrics are being used. The optimum will depend upon the drapability of the fabric, its ability to stack in successive layers and the compressibility of the fabric.

Higher fibre volume fractions and levels of properties may be achieved by filament winding of individual rovings (Table 4.8, section 11.7) or by using aligned prepregs (e.g. Table 5.17), if a higher level of properties is required.

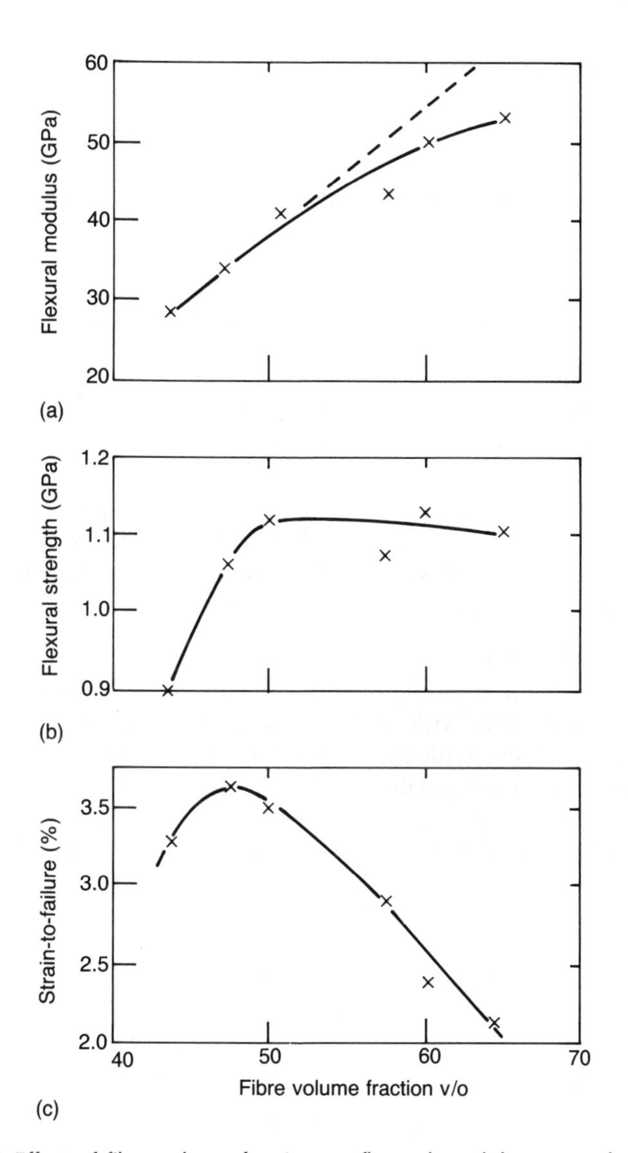

Figure 4.8 Effect of fibre volume fraction on flexural modulus, strength and strain-to-failure (Chianumba).

Source

A. Chianumba at Reading University.

■ 4.2 POLYESTER RESINS

■ **Temperature**

Mechanical properties

Effect of temperature

Materials

The *materials* are:

- glass – chopped strand mat, M4 450 g/m^2 (Vetrotex);
- resin – ortho-polyester, Norpol 44 (Jotun Polymer).

Manufacture

Manufacture was achieved by contact moulding and hand lay-up.

Observations

At low temperatures (down to at least −60°C), glass/polyester laminates get stronger as well as stiffer, and there is little evidence of embrittlement.

The upper working limit is set by the heat deflection temperature of the resin (62°C for this resin) with a drop in modulus of 40% at this temperature. The tensile strength decreases more slowly than the modulus with increasing temperature.

Design implications

Composites made from thermosetting resins like polyesters can be used at low temperatures without becoming embrittled. This is much more difficult to achieve with thermoplastics. The maximum working temperature of a resin is usually 20°C below the HDT (Chapter 10).

Higher temperature versions of the principal types of polyester resin (Table 4.1) are available and these will have a similar variation of property with temperature. Iso-polyesters have the highest HDT (*c.*130°C) with iso-NPG polyesters somewhat lower (*c.*110°C) and ortho-polyesters the lowest (*c.*100°C).

A comparison of the high temperature capability of the four principal thermosetting resins is given in Figure 4.14.

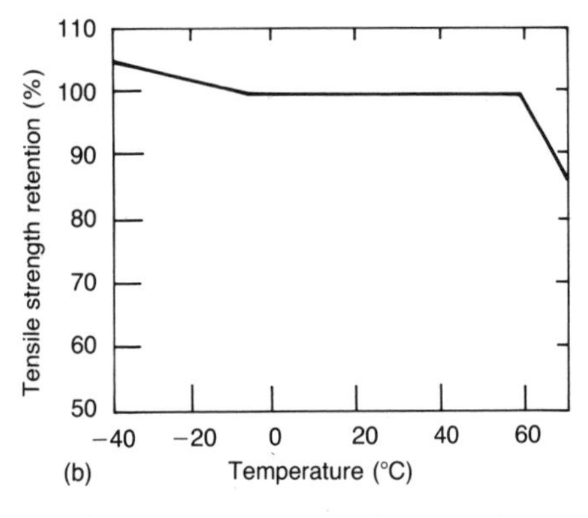

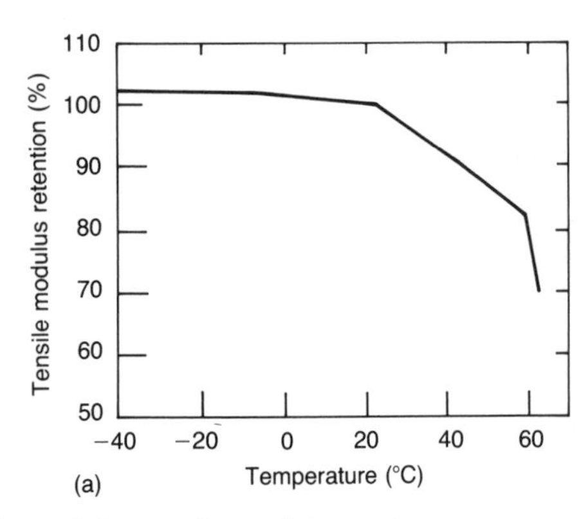

Figure 4.9 Variation of (a) tensile strength, and (b) tensile modulus with temperature (Arvesen).

Source

R. Arvesen at Jotun Polymer.

4.3 VINYL ESTER RESINS

Properties of cast resins

Typical properties of cast vinyl ester resins are listed in Table 4.5.

Materials
The *materials* are:

- flexible resin – Derakane 411 (Dow Chemicals);
- high temperature resin – Derakane 470;
- rubber-modified resin – Derakane 510.

Manufacture
Manufacture was achieved by contact moulding.

Observations
The vinyl esters can, like polyesters, be divided into three groups on the basis of either elongation to break or load capability at high temperature.

For these resins, a high heat distortion temperature goes with low elongation and vice versa.

Design implications
These resins are generally regarded as having properties between those of polyesters (Table 4.1) and epoxies (Table 4.9). This applies both to mechanical and chemical properties as well as the ease of use with various manufacturing processes.

However, the cost of vinyl esters lies between those of polyesters and epoxies so advantages in properties have to be traded off against cost.

Table 4.5 Typical properties of cast vinyl ester resins

		flexible	high temperature	rubber modified
viscosity	mPa.s	500	200	350
density	Mg/m^3	1.12	1.15	1.13
tensile strength	MPa	81	73	70
tensile modulus	GPa	3.3	3.5	3.1
elongation at break	%	5	3–4	8
flexural strength	MPa	124	133	135
flexural modulus	GPa	3.1	3.8	3.2
Barcol hardness		35	40	40
heat distortion temperature	°C	102	145	80
compressive strength	MPa	114	127	86
compressive modulus	GPa	2.4	2.2	2.2

Source
The Dow Chemical Company.

■ 4.3 VINYL ESTER RESINS

■ Comparison with polyester

Effect of resin type

A comparison of the tensile and compressive properties of polyester and vinyl ester composites is given in Table 4.6 for a combination fabric and in Table 4.10, and 4.11 for other mats and fabrics.

Materials
The *materials* are:

- glass – balanced warp/weft woven fabric 800 g/m^2, chopped strand mat 100 g/m^2 Rovimat (Chromarat);
- resins – ortho-polyester, Norpol 41-90 (Jotun Polymer); iso-polyester, Norpol 72-80 (Jotun Polymer); iso-NPG-polyester, Norpol 20-80 (Jotun Polymer); flexible vinyl ester, Norpol 92-20 (Jotun Polmyer); rubber-modified vinyl ester, Norpol 92-40 (Jotun Polymer).

Manufacture
Manufacture was achieved by contact moulding through hand lay-up.

Observations
There is little variation in tensile properties between the various composites, but a greater variation in compressive properties, particularly the strength.

The compressive properties are greater than or equal to the tensile properties for all laminates. This is unusual as generally compressive properties are similar to or lower than tensile properties. The failure strain of the composites is remarkably similar even though the failure strain of the matrices varies from 4.6 to 8% (Tables 4.1, 4.5).

Design implications
The mechanical properties displayed in Table 4.6 are determined by the reinforcement rather than the resin type. The data in Tables 4.11 and 4.12 confirm this supposition for both chopped strand mat and woven fabrics.

Thus the choice of resin is governed by other considerations such as cost, processability or properties such as temperature capability, environmental resistance or impact.

Whilst with these composites the compressive properties are higher than the tensile properties, this is not always valid, particularly for aramid (Table 5.7) and carbon composites (Table 5.14).

The properties of composites based on woven fabric and chopped strand mat and different matrices are compared in Tables 4.10 and 4.11.

4.3 VINYL ESTER RESINS

Comparison with polyester

Table 4.6 Properties of vinyl ester and polyester composites

		polyester			vinyl ester	
		ortho-	iso-	iso-NPG	flexible	rubber-modified
glass content	v/o	53	53	51	53	52
tensile						
strength	MPa	226	209	205	227	237
modulus	GPa	17.3	16.9	16.1	16.8	16.9
failure strain	%	1.8	1.4	1.6	1.7	1.7
compressive						
strength	MPa	294	209	258	288	259
modulus	GPa	17.5	17.5	16.5	16.6	15.9
failure strain	%	1.9	1.4	1.6		1.7

Source

Echtermeyer, A.T. *et al.* (1991) Significance of damage caused by fatigue on mechanical properties of composite laminates, *Proc. 8th International Conference on Composite Materials*, Hawaii, Paper B1.

■ 4.3 VINYL ESTER RESINS

■ **Temperature**

Mechanical properties

Effect of temperature

Materials
The *materials* are:

- glass;
- resins – vinyl ester, flexible – Derakane 411-45 (Dow Chemicals), high temperature – Derakane 470-36 (Dow Chemicals), rubber-modified – Derakane 510-A40 (Dow Chemicals).

Manufacture
Manufacture was achieved by contact moulding.

Observations
The data show the working temperature range above 0°C over which vinyl esters can operate. The upper working limit is set by the HDT (Table 4.5) with typically a reduction of *c*.30% at this temperature. These resins also exhibit a high resistance to thermal ageing when used for extended periods at elevated temperatures.

Design implications
Vinyl esters have good chemical resistance (Table 9.6) as well as good temperature resistance so this makes them very attractive for applications such as chemical vessels, tanks and pipes. However, properties must be checked by undertaking suitable accelerated or long-term tests (section 10.3).

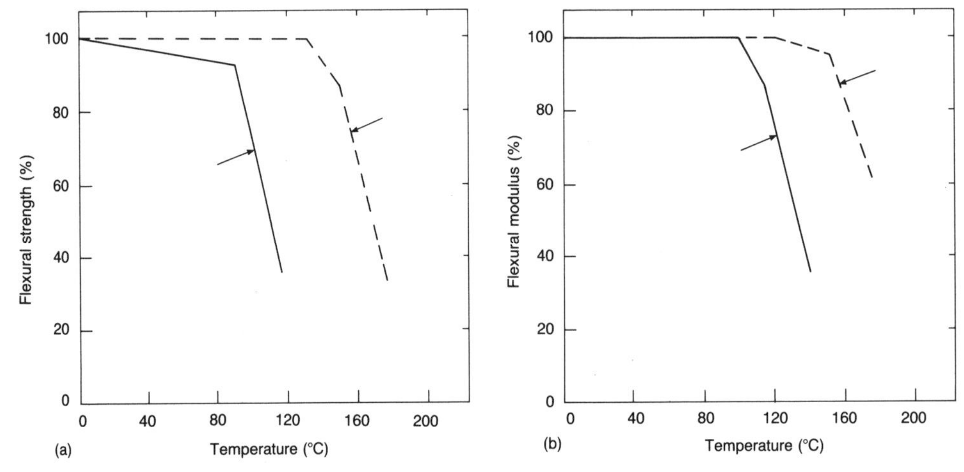

Figure 4.10 Variation in (a) flexural strength, and (b) flexural modulus with temperature for glass-reinforced vinyl ester composite (Dow Chemicals).

Source
The Dow Chemical Company.

4.4 PHENOLIC RESINS

Properties of cast resins

Phenolic resins exist in both novalac and resole form. The latter are able to be cured like polyesters with the addition of catalysts and/or heat and so are of greatest interest for general purpose engineering. The available information covers resins cured with acid catalysts and this is assumed unless otherwise stated.

Typical properties of a cast resin are listed in Table 4.7.

Observations

The viscosity range selected depends upon manufacturing process and application. These resins are cold cured by adding a liquid (acid-based) catalyst. Though cure is possible at 20°C, it is preferable to post cure in the range 60°–80°C to obtain dimensionally stable components.

Design implications

For large mouldings, low shrinkage, which is characteristic of this resin, is an important design consideration. The high heat distortion temperature combined with excellent fire resistance (section 8.6) has allowed these materials to become more widely used.

Table 4.7 Typical properties of cast resin

viscosity range	mPa.s	200–300; 500–600; 1500–1700; thixotropic
density	Mg/m^3	1.24
tensile strength	MPa	24–40
tensile modulus	GPa	1.5–2.5
flexural modulus	GPa	3.0–4.0
flexural strength	MPa	60–80
elongation at break	%	1.8
heat distortion temperature	°C	*c.*250
thermal expansion	$°C^{-1}$	$10–20 \times 10^{-6}$
thermal conductivity	W/m.K	0.2–0.25
in-mould shrinkage	%	0.5 in plane; 3–5 out of plane
electrical resistivity	Ω.m	3×10^9
dielectric strength	kV/mm	5.6
dielectric loss	–	1.4 (200 Hz); 0.3 (1 kHz); 0.05 (1 MHz)
water absorption	%	0.4
critical oxygen index	%	45–55

Source

M. Orpin at BP Chemicals.

■ 4.4 PHENOLIC RESINS

■ **Chopped strand mats**

Flexural strength and binder

Effect of chopped strand mat binder

The effect of binder used to hold together the chopped rovings to form a chopped strand mat has been investigated with phenolic resin (Figure 4.11).

Materials

The *materials* are:

- glass – chopped strand mats, one emulsion bonded, three powder bonded;
- resin – phenolic (CEAT).

Manufacture

Manufacture was achieved by contact moulding through hand lay-up; additional time allowed for good wet-out.

Observations

A wide range of properties is attainable with a similar variation to that obtained with polyester resins (Table 4.2). The binder needs to be dissolvable in phenolic resin, but most mats are designated for use with polyester resin, and so may not be ideal for phenolic resin.

The variation in impact strength is similar to that of flexural strength.

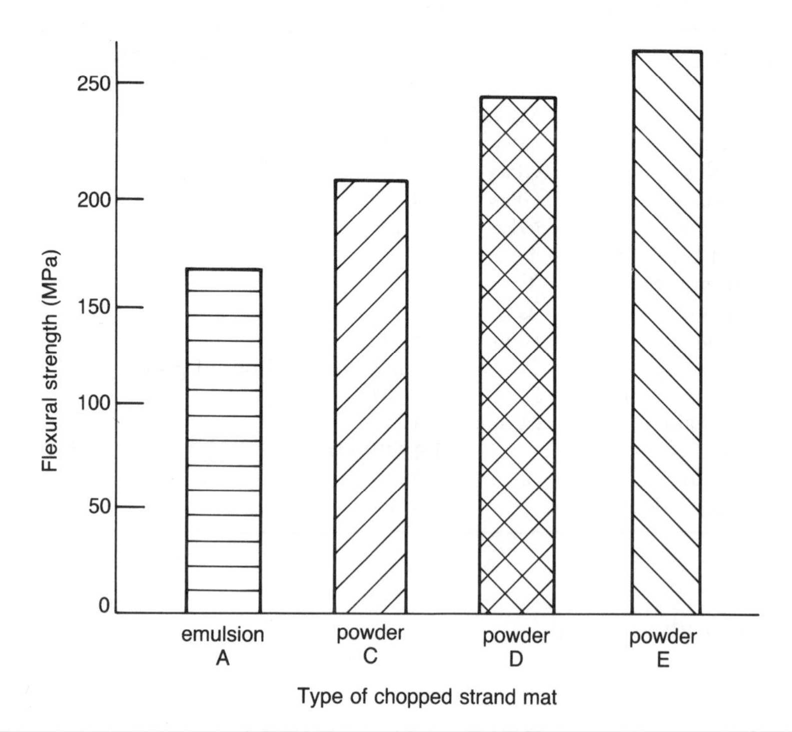

Figure 4.11 Comparison of flexural strengths of CSM laminates with various binders (Sundaram).

4.4 PHENOLIC RESINS

Chopped strand mats

Flexural strength and binder

Design implications

The type of binder is clearly important if a desired level of properties is required for a specified thickness. Whilst these results relate to short-term properties, the need for good fibre/resin compatibility is even more important for long-term properties, including environmental protection and creep/fatigue (Figure 10.16).

Source

Sundaram, S. (1990) Development of phenolic compatible glass reinforcements for the growing Indian market. *Popular Plastic Packaging*, 35, 17–22.

■ 4.4 PHENOLIC RESINS

■ **Shear strength**

Reinforcement type

Interlaminar shear strength (ILSS)

Materials

The *materials* are:

- glass – chopped strand mats, emulsion bonded (E1/E2), powder bonded (P3/P4), balanced woven roving fabric (WR), unidirectional rovings (UD);
- resin – phenolic J2018L (BP Chemicals).

Manufacture

Manufacture was achieved by resin transfer moulding using cold cure catalyst.

Observations

Chopped strand mats show a wide variation in ILSS; composites having a low flexural strength (Figure 4.11) also exhibit a low shear strength. The woven fabric and unidirectional fabrics have higher flexural strengths (Table 4.8) and higher ILSS (Figure 4.12).

Design implications

High values of ILSS will generally provide a correspondingly high level of properties in other directions. Tests other than ILSS are available to check fibre/resin compatibility and designers should clarify the most appropriate glass type with the material supplier.

ILSS clearly depends upon the nature of the reinforcement and the

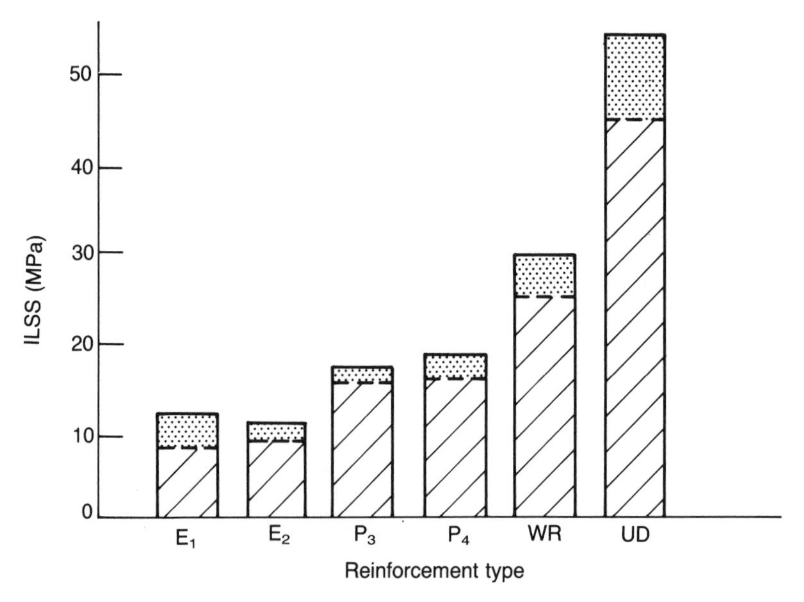

Figure 4.12 Interlaminar shear strength of phenolic laminates with different reinforcements; data scatter is indicated by broken line (Sundaram, Tavakoli and Orpin).

Shear strength

Reinforcement type

type of binder (finish). Various studies indicate that the use of glass with a silane finish is best.

Sources

M. Orpin at BP Chemicals.

Sundaram, S. (1990) Development of phenolic compatible glass reinforcements for the growing Indian market. *Popular Plastic Packaging*, 35, 17–22.

Tavakoli, S. *et al.* (1989) Compatibility of commercial CSM with a phenolic resin system for hand laminating. *Composites*, 20, 159–65.

■ 4.4 PHENOLIC RESINS

■ **Reinforcement type**

Mechanical properties

Reinforcement type

Typical properties of laminates made with various types of reinforcement are listed in Table 4.8.

Materials

The *materials* are:

- glass – chopped strand mat (OCF MK 12), continuous filament mat, woven fabric 800 g/m^2, unidirectional (Silenka 051L);
- resin – phenolic, J2027L (BP Chemicals).

Manufacture

Manufacture was achieved through either contact moulding by hand lay-up, or filament winding of unidirectional rovings.

Observations

With mats and fabrics suitably sized for phenolic resins, mechanical properties are similar to those of more widely used laminating resins. The moulders have adjusted the process conditions to obtain good wet-out and hence good properties.

Design implications

These results suggest that for the initial design stages, similar levels of properties can be assumed for the main thermosetting laminating resins. However the interlaminar shear strength of phenolic composite can be appreciably less than polyester or epoxy ones and so the level of shear stresses may need to be checked.

Table 4.8 Laminates with various reinforcements

		CFM	CSM	woven fabric	uni-directional
glass content	v/o	15	24	42	63
tensile strength	MPa	40–60	100–150	280–350	
tensile modulus	GPa	4.5–6.0	5.5–7.5	16–20	
elongation at break	%	1.0–2.0	1.8–2.5	2.0–2.5	2.9
flexural strength	MPa	75–125	150–230	300–450	1340
flexural modulus	GPa	4.0–5.5	5.7–7.5	17–20	46

Sources

Forsdyke, K.L. (1984) Phenolic resin composites for fire and high temperature applications, in *Fibre Reinforced Composites*, PRI, London.

Sundaram, S. (1990) Development of phenolic compatible glass reinforcements for the growing Indian market. *Popular Plastic Packaging*, 35, 17–22.

4.4 PHENOLIC RESINS

High-temperature soaking

Flexural strength

Soaking at elevated temperatures

Materials

The *materials* are:

- glass – Equerove EC17 (Pilkington);
- carbon – Grafil EXAS (Courtaulds);
- resins – phenolic, J2018L (BP Chemicals).

Manufacture

Manufacture was achieved by unidirectional winding of rovings onto a plate.

Observations

No reduction, a possible increase, in flexural strength for carbon/phenolic. The good bond between phenolic resin and the epoxy size on carbon fibres could be a contributory reason.

An initial drop in strength is a feature of some glass/phenolic systems with a loss of up to 30% flexural strength in the first hour at 200°C. Thereafter no further loss in strength is detected over an extended time period.

Design implications

Phenolics clearly have a high temperature capability, which opens up specific applications. The mode of use should be checked with the material supplier as regards environmental resistance.

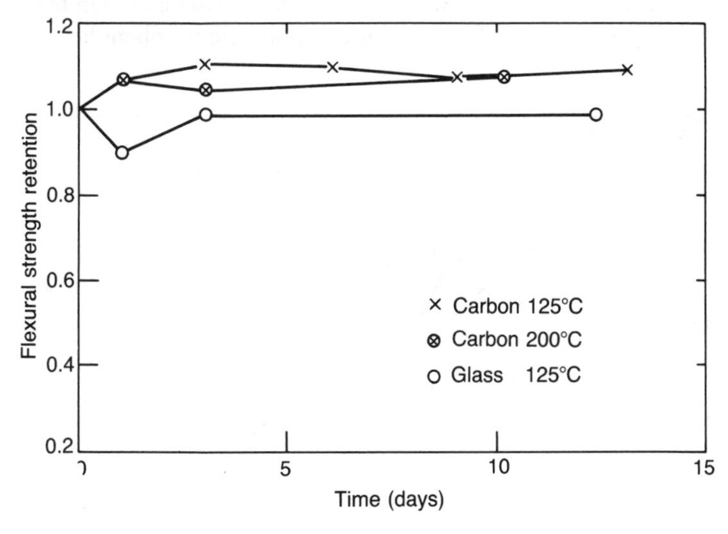

Figure 4.13 Effect of soaking at elevated temperatures on flexural strength of glass and carbon reinforced phenolic laminates normalized to strength at zero time (Forsdyke).

Sources

Forsdyke, K.L. (1984) Phenolic resin composites for fire and high temperature applications, in *Fibre Reinforced Composites*, PRI, London.

Forsdyke, K.L. (1988) Phenolic matrix resins – the way to safer reinforced plastics, *Proc. 43rd Conference of the SPI*, session 18 C/l.

■ 4.4 PHENOLIC RESINS

■ **Other resins**

Temperature

High temperature capability of resins

Materials
The *materials* are:

- glass – Equerove EC17 (Pilkingtons);
- resins – iso-polyester, 73.2661 (DSM); epoxy, MY 750/HT 972 (Ciba-Geigy); phenolic, J2018L (BP Chemicals); vinyl ester, Derakane 470 (Dow Chemicals).

Manufacture
Manufacture was achieved through unidirectional filament winding in the form of a plate.

Observations
These four thermosetting resins have somewhat different load-carrying capabilities at elevated temperatures. Polyester, epoxy and vinyl ester have similar heat distortion temperatures (130–150°C) and from these data, one can infer that epoxy is best at intermediate temperatures and vinyl ester at higher temperatures. Overall, however, the phenolic has the best temperature resistance, in line with its much higher HDT (250°C).

Design implications
Time, temperature and load are important in assessing high temperature capability of a resin. For structural applications, phenolics have clear

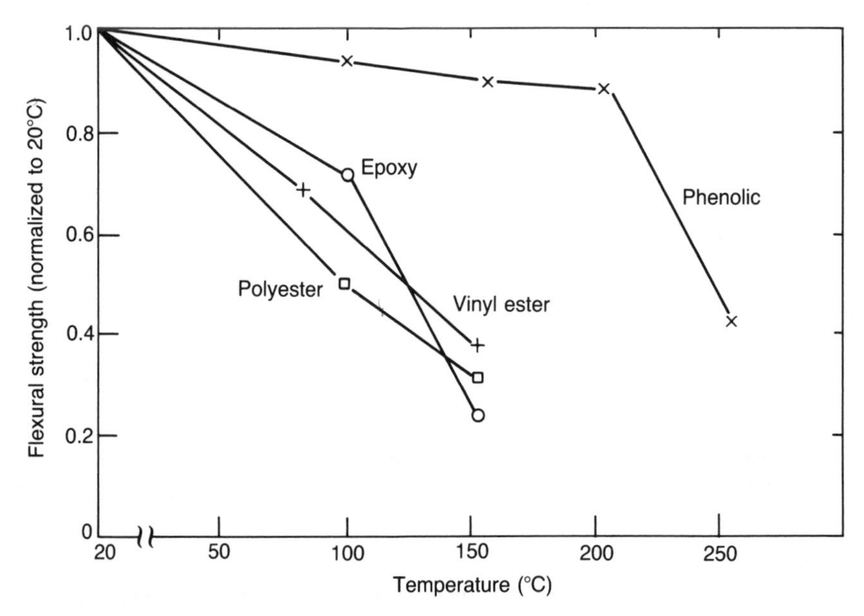

Figure 4.14 Flexural strength of glass reinforced composites at elevated temperatures after soaking for 30 minutes (Forsdyke).

advantages over the other thermosetting resins above. So one must assess the likelihood of a high temperature excursion even if it is only for a short time.

Other resins

Temperature

There are very specialized thermosets and thermoplastics, which have a capability up to 300°C (Tables 3.2, 5.21), but are little used outside the aerospace industry, largely because of their cost.

Source

Forsdyke, K.L. (1988) Phenolic matrix resins – the way to safer reinforced plastics, *Proc. 43rd Conference of the SPI*, session 18 C/1.

■ 4.5 EPOXY RESINS

■ **Cast resins**

Properties

4.5 EPOXY RESINS

Epoxy resins are a versatile group of materials widely used in composite applications. Epoxy is the general classification for resins containing two carbon and one oxygen atom bonded in a ring. Such resins may be derived from many different starting materials such as phenol, bisphenol and multi-functional phenolic.

Epoxies are used in combination with a number of co-reactants or hardeners and catalysts to give a wide range of properties.

Two basic forms are used for composites. A di-functional form is used for general purpose applications, which may be either cured with an anhydride or an amine. The use of aliphatic amines as accelerators/ catalysts allows cure to occur at low temperatures. The tetra-functional form is generally used for aerospace applications often requiring high temperatures, and is generally cured with amines.

Typical properties of two cast resins are given in Table 4.9.

Table 4.9 Typical properties of cast epoxy resins

		cold curing	hot curing
viscosity	mPa.s	600–1000	200–400 (at 60°C)
density	Mg/m^3	1.1	1.1
tensile strength	MPa	60–80	70–85
tensile modulus	GPa	3.8–3.0	3.0–2.6
elongation at break	%	1.9–7.5	4.5–8.0
Poisson's ratio		0.35	0.35
shear modulus	GPa	1.1	1.0
HDT	°C	50–90†	165

† HDT depends on whether component is post-cured.

Materials

The *resins* are:

- cold curing LY 564/HY 564 (Ciba Geigy);
- hot curing LY 556/HT 972 (Ciba Geigy).

Note: the form of the notation used here is resin type/hardener, so that in the case of cold curing LY 564 is the resin type, HY 564 is the hardener.

Comparison of mats

The properties of chopped strand mat in a typical epoxy resin is contrasted in Table 4.10 and Figure 4.15 with data obtained with other resins.

Materials
The *materials* are:

- reinforcement – CSM 450 g/m^2;
- resins – urethane-modifed vinyl ester (Scott-Bader); iso-polyester (Scott-Bader); 60:40 blend urethane acrylate/polyester (Scott-Bader); iso-polyester (high V_f) NI423 (Cdf Chemie).

Manufacture
Manufacture was achieved through contact moulding; post curing 16 hours at 40°C (equivalent to 28 days at 20°C).

Table 4.10 Properties of CSM laminates in various resins

		epoxy	urethane vinyl-ester	iso-polyester	60:40 urethane/ polyester	iso-polyester*
glass content	v/o	17	18	20	17	30
tensile						
strength	MPa	122	122	129	124	170
modulus	GPa	8.1	6.7	8.8	6.6	12.2
elongation at break	%	1.5	1.8	1.5	1.9	
flexural						
strength	MPa	199	224	243	192	280
modulus	GPa	6.9	7.4	7.6	5.5	11.8
elongation at break	%	2.9	3.0	3.2	3.5	
compressive strength	MPa					220
shear strength	MPa	12	10	9	11	

* Mid-point of data as measured by Creux.

Observations
These results show that there are small differences between the various resin systems with CSM as the reinforcement. The urethane-modified vinyl ester and polyester are more flexible resins, which are likely to lead to better impact properties (section 6.1) under the post curing conditions used.

■ 4.5 EPOXY RESINS

■ **Chopped strand mats**

Mechanical properties

Materials

The *materials* are:

- reinforcement – CSM 500 g/m^2 (Vetrotex);
- resins – polyester NI423 (CdF Chemie); epoxy LY556/HY951 (Ciba-Geigy); vinyl ester D411-45 (Dow Chemicals).

Manufacture

Manufacture was achieved through contact moulding, hand lay-up.

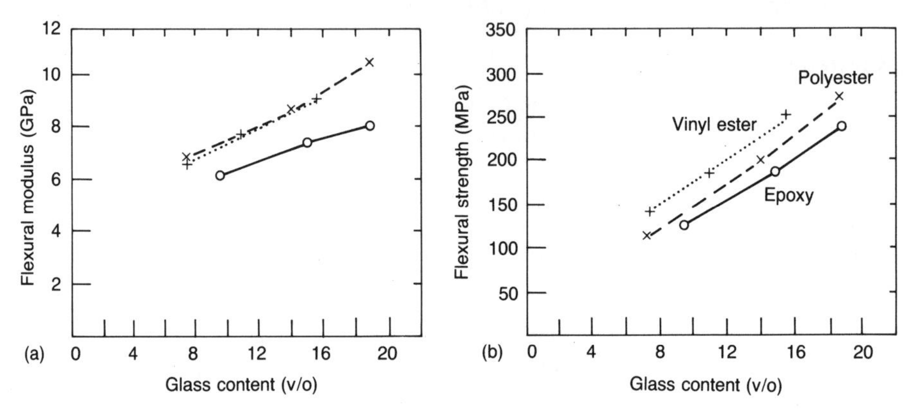

Figure 4.15 (a) Flexural modulus, and (b) flexural strength of CSM laminates with various resins (Creux).

Design implications

For applications with low glass content, the choice of resin is more likely to depend upon cost (polyester being the cheapest), processability or chemical/environmental resistance (Chapter 9).

Sources

L. Norwood at Scott-Bader.

G. Creux at Vetrotex.

Properties of woven fabrics

The properties of balanced woven fabrics in an epoxy resin are listed in Table 4.11 together with those obtained with some other thermosetting matrices.

Materials

The *materials* are:

- resins – as Table 4.10;
- reinforcement – plain weave fabric.

Manufacture

Manufacture was achieved through contact moulding; post cure 16 hours at 40°C.

Observations

As with chopped strand mats (Table 4.10), there are only minor differences between the four composite systems for these properties, glass/epoxy having the highest shear strength. Observe that the tensile and compressive strengths are lower than the flexural strengths.

Design implications

Unless shear strength is important, considerations other than simply strength and modulus along the fibre directions will dictate the choice of resin system. For high fibre volume fractions (>40%) fine fabrics might be needed.

Table 4.11 Typical properties of woven fabric laminates

		epoxy	urethane vinyl-ester	iso-polyester	60:40 urethane/ polyester	iso-polyester*	
glass content	v/o	35	38	35	35	40	50
tensile							
strength	MPa	212	243	241	225	320	440
modulus	GPa	16	16	16	13	20	25
elongation at break	%	1.3	1.5	1.5	1.7	1.6	1.7
flexural							
strength	MPa	365	372	366	334	480	590
modulus	GPa	19	13	13	12	17	22
elongation at break	%	2.7	2.8	2.9	2.9	3.0	2.7
compressive strength	MPa					260	320
shear strength	MPa	11	9	7	7		

* Mid-point of data as measured by Creux.

Sources

L. Norwood at Scott-Bader.
G. Creux at Vetrotex.

■ 4.5 EPOXY RESINS

■ **Fine fabrics**

Mechanical properties

Properties of fine fabrics

Higher levels of properties can be obtained with fine fabrics than with coarse ones (say, 150–250 g/m^2 compared with 500–1000 g/m^2). Some typical data are given in Table 4.12.

Materials

The *materials* are:

- resins – cold curing LY556/HT972 (Ciba-Geigy); hot curing LY564/HY964 (Ciba-Geigy);
- reinforcement – (a) 16 layers, 91745 (Interglas); (b) 12 layers, 92146 (Interglas); (c) OCF 859, woven rovings.

Manufacture

Manufacturing was achieved through contact moulding by hand lay-up.

Observations

With fine fabrics, a high fibre volume fraction can be obtained, which will give a proportionately higher level of properties. The properties of the cold cure laminate can be improved by post curing at 50°C. Each fabric gives a layer thickness of *c*.0.25 mm.

Table 4.12 Typical properties of fine fabric epoxy laminates

		cold curing epoxy	hot curing epoxy	reinforcement
glass content	v/o	48	50	
(a) flexural				
strength	MPa	440–455	520–570	16 layers
modulus	GPa	21.9–22.5	19.3–19.9	Interglas 91745 fabric
elongation at break	%	2.7–3.0	3.1–3.4	
glass content	v/o	65	65	
(b) interlaminar shear				12 layers Interglas 92146 fabric
strength	MPa	42–47	55–59	
(c) transverse tensile				
strength	MPa	40–44	45–51	OCF 859 woven
modulus	GPa	19.5–21.0	17.2–19.0	rovings
elongation at break	%	2.0–2.2	3.0–3.3	
shear				
strength	MPa	55–57	67–71	
modulus	GPa	7.0–8.5	5.3–6.8	
compressive				
strength	MPa	120–130	135–140	
modulus	GPa	18.0–19.5	15.0–17.0	

Design implications
Fabrics provide a wide range of properties and are chosen for both ease of drape (fine being better than coarse) and fibre orientation. Prepreg systems are available with fabrics as well as fully aligned rovings.

Source
Ciba-Geigy trade literature.

■ 4.5 EPOXY RESINS

■ **Prepreg fine fabrics**

Temperature effects

Effect of temperature

Fine fabrics are also available in prepreg form and these fabrics are typically carbon or glass or some other hybrid fabric. Typical properties are listed in Table 4.13 for a fine carbon fabric.

Materials

The *materials* are:

- reinforcement – carbon fabric, satin weave, 285 g/m^2, 60 v/o;
- resin – prepreg Fibredux 914C-833-40 (Ciba-Geigy).

Manufacture

Manufacture was achieved by prepreg moulding followed by a post cure regime of 1 hour at 175°C, then 4 hours at 190°C.

Observations

The moduli are all very high due to the reinforcement being carbon rather than glass (compare moduli in Table 3.1). The strengths are somewhat higher than those achievable with glass (compare with Table 4.12).

The property retention across the temperature range −50°C to +200°C is shown in Figure 4.16.

There is a slight decrease in both tensile and flexural strength below 0°C (*c*.7% and 12%), but the tensile modulus is hardly altered.

The tensile strength is hardly affected by elevated temperatures up to 200°C through use of a suitable high temperature matrix system. However, the interlaminar shear strength drops steadily and this is reflected in the decrease in compressive strength and therefore in the flexural strength.

Design implications

Prepregs, though more expensive, tend to give high and reliable properties as the resin/reinforcement process route has been optimized. Such

Table 4.13 Typical properties of a fine carbon/epoxy prepreg at room temperature

		warp	weft
tensile strength	MPa	625	630
tensile modulus	GPa	69	65
flexural strength	MPa	903	964
flexural modulus	GPa	60	56
compressive strength	MPa	667	595
compressive modulus	GPa	87	82
interlaminar shear strength	MPa	63	57
short beam shear	MPa	65	63

4.5 EPOXY RESINS ■

Prepreg fine fabrics ■

Temperature effects

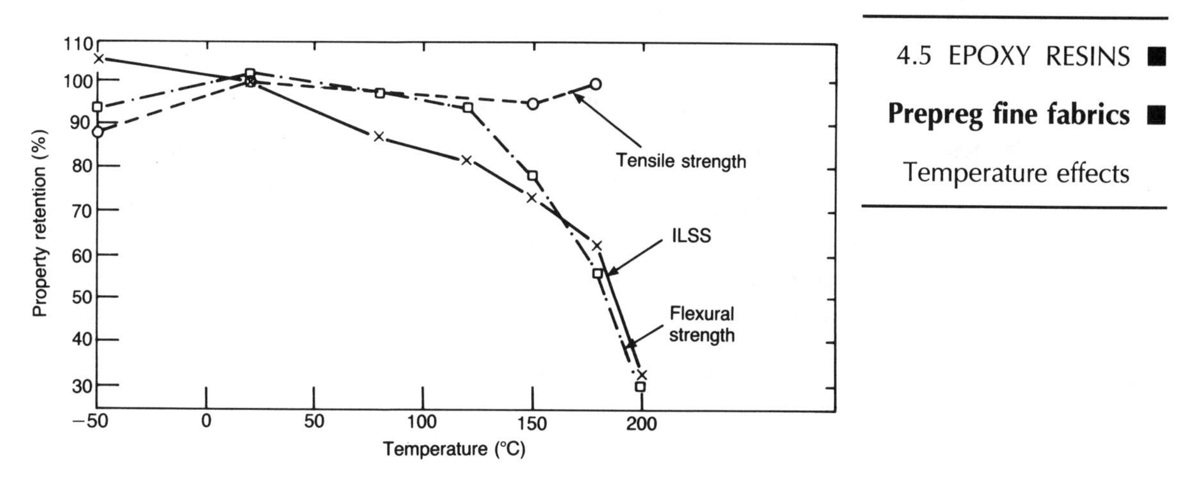

Figure 4.16 Strength retention as a function of temperature for a carbon epoxy prepreg (Ciba-Geigy).

materials are generally used where performance rather than cost is an important consideration, for example, in aerospace applications.

Epoxy prepregs are numerous and form the backbone of modern composite applications in aircraft.

Other prepreg systems are described in Chapter 5. Because of the cost of the fibre, carbon fabrics generally tend to be fine and are not generally used in random mat form like glass.

Source

Ciba-Geigy trade literature.

■ 4.6 POLYPROPYLENE

■ **Glass mat thermoplastics**

4.6 OTHER RESINS

Glass mat thermoplastics (GMT)

Thermoplastic resins are generally reinforced with short fibres (for use with injection moulding and not considered in this book), with long fibres (thermoplastic sheet compound, Chapter 5) or with glass mat (i.e. glass mat thermoplastics or GMT). Polypropylene is the resin most used as the matrix for GMT and the reinforcement is usually a random mat, primarily chopped strand but it may be continuous filament or needled mat.

Some typical data are given in Table 4.14.

Materials
The *materials* are:

- reinforcement – glass mat RB34BXI (OCF);
- resin – polypropylene.

Manufacture
Manufacture is achieved by hot press moulding.

Observations
These materials compete with sheet moulding compounds (SMC) for applications which are generally of high volume (section 11.4). Compared with SMC (Table 11.11) they have a lower modulus, and a higher failure strain.

Design implications
These materials have only recently become available in Europe and development work is still under way to develop suitable glass sizings and reduce the melt viscosity (by adding a small amount of plasticizer to the resin). Unlike SMC, these materials have so far only been used for semi-structural applications, primarily motor vehicles.

Table 4.14 Properties of glass mat thermoplastics

		OCF	Azdel	Azdel
glass content	w/o	29.6	30	40
specific gravity	Mg/m^3		1.13	1.19
flexural strength	MPa	121–131	140	160
flexural modulus	GPa	5.9–6.6	4.5	5.5
tensile strength	MPa	81–77	85	95
tensile modulus	GPa	5.8–5.5	5.5	7.0
tensile elongation	%	2.9–3.1	2	2
heat distortion temperature	°C	150	165	165
mould shrinkage	%		0.2–0.3	0.2–0.3

Sources

Reinforcements for polypropylene: technology developments. *Reinforced Plastics*, March 1992, 18–21.

Technopolymers challenge metals. *Reinforced Plastics*, March 1992, 26–7.

4.6 ACRYLATES ■

Cast resins ■

Properties

Properties of cast resins

A range of methacrylate resins has been developed. These resins have a low viscosity, high reactivity and good resin toughness. They are thermosets which can be cold cured with a range of catalysts and accelerators to give flexibility in use with various moulding processes.

Typical properties of cast acrylic resins are given in Table 4.15.

Table 4.15 Typical properties of cast acrylic resins

		835S	826HT
viscosity	mP.s	60	135
density	Mg/m^3	1.08	1.12
tensile strength	MPa	56	62
tensile modulus	GPa	2.1	3.2
strain-to-failure	%	6.5	2.6
flexural strength	MPa	132	127
flexural modulus	GPa	2.9	4.8
heat distortion temperature	°C	83	123
in-mould shrinkage	%	11–12	
Barcol hardness		40	55
impact strength	kJ/m^2	30	17

Materials

The *materials* are:

- Modar 835S (ICI);
- general purpose grade Modar 826HT (ICI), pultrusion grade.

Observations

The ultimate strengths are similar to those of polyesters (Table 4.1). The impact strength and strain to failure indicate that these are tough resins. In-tension failure is initiated by yielding rather than brittle fracture. 826HT is a low shrinkage resin developed especially for pultrusion.

Design implications

With a suitable catalyst a fast moulding time, 2–5 minutes, can be attained. The low viscosity allows a high filler loading. A possible filler, for fire resistance, is aluminium trihydrate (Figure 8.4 and Table 8.5).

Source

ICI Acrylates, trade literature.

4.7 DESIGN STRATEGY

In this chapter we have considered the properties of composites reinforced with mats or fabrics rather than aligned rovings (Chapter 5). The selection process is set out schematically in Figure 4.17.

As set out in this chapter, the property level increases as one proceeds from chopped rovings through random mats to fabrics, but so does the cost and the difficulty in conforming to complex curvature.

Likewise the modulus increases as one moves from glass to aramid and

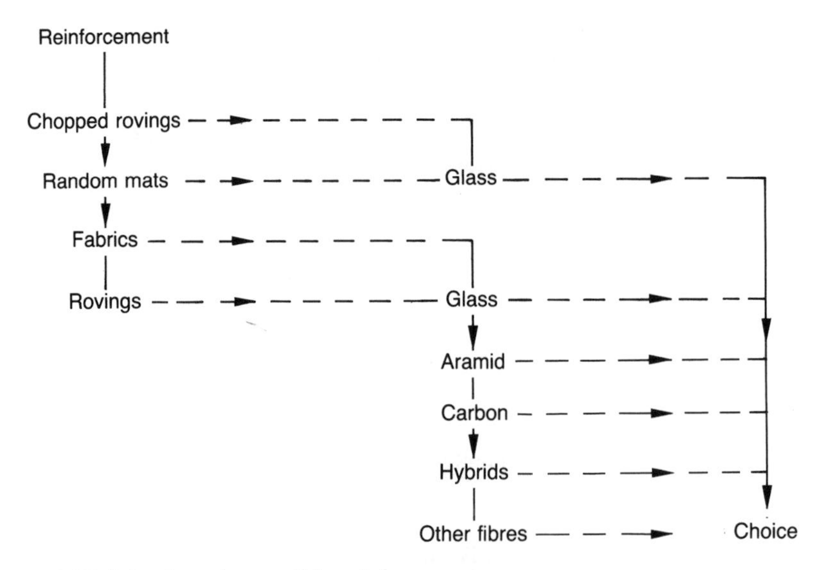

Figure 4.17 Selection of a possible reinforcement.

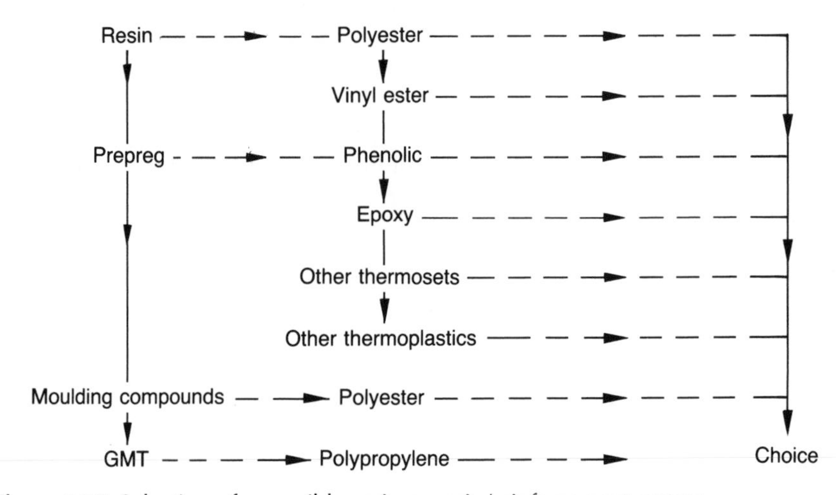

Figure 4.18 Selection of a possible resin or resin/reinforcement system.

then to carbon, but so does the cost, whilst considerations of other properties such as impact (Chapter 6) may dictate the use of fabrics with hybrid reinforcement (like glass/carbon).

The resin selection process is set out in Figure 4.18.

With the matrix, the neat resin provides the widest choice of properties followed by prepregs and moulding compounds (section 11.4), with GMT offering the least choice (at present).

The advantage of knowing the process route and property level is attractive for many applications, but this has to be offset by either cost (prepregs) or tooling requirements (moulding compounds and GMT).

For the properties and resins discussed in this chapter, a range of thermosetting matrices is available and the choice may be dictated by either cost (section 2.3) or balance of properties (as discussed in other chapters).

4.8 REFERENCE INFORMATION

General

Weatherhead, R.G. (1980) *FRP Technology*, Elsevier, Barking – comprehensive description of resins.

Engineered Materials Handbook, Volume 1: Composites (1987) ASM International, Metals Park, Ohio – data in form suitable for design purposes.

Nordic Boat Standard for craft under 15 metres (1990) Det Norske Veritas, Hovik. Similar standards are set by other classification societies such as Lloyd's Register and Germanischer Lloyd.

Update on GMT in *Reinforced Plastics*, March 1992 – series of articles on current status of GMT and how it is being used.

Materials suppliers

These need to be consulted to select a suitable reinforcement and resin once the initial selection has been made.

To help select the type of fabric, consult Mayer, R.M. (ed.) (1993) *Design with Reinforced Plastics*, Ch. 5, Design Council, London.

5 MECHANICAL PROPERTIES OF ALIGNED FIBRE COMPOSITES

SUMMARY

This chapter is concerned with the short-term mechanical properties – moduli and strengths – of glass, aramid and carbon fibres in a thermosetting resin matrix. A little information on reinforced thermoplastic matrix systems is also included. The data mainly refer to the room temperature properties of 55–65 v/o fibre, unidirectional, systems. The effects of the variation in fibre volume loading, method of test and instantaneous and long term exposure to temperature are briefly mentioned. Longitudinal properties tend to be fibre dominated, and so are compressive properties to some extent for glass and carbon fibres. The anisotropy of unidirectional materials is noticeable.

5.1 GENERAL INFORMATION

Although glass fibre-reinforced composites were developed in the 1940s it was, arguably, the introduction of the very stiff, strong, low density carbon and aramid fibres in the 1960s that led to the present composites industry. These fibres, with a wide range of properties, enabled structures with high specific properties (the ratio of the property to the specific gravity) and excellent thermal stability to be designed and built. In order to optimize the performance of the composite, new resins were developed and existing systems refined, while the advances in the manufacturing methods, design procedures and the understanding of the basic behaviour of the fibre-reinforced materials improved the performance of existing composites based on glass fibres and polyester or phenolic resins.

5.2 GLASS FIBRES

The most extensively used reinforcement, for both thermosetting and thermoplastic matrices, is glass fibre. This may be in the form of milled, chopped, random mat, woven or unidirectional material, though here only unidirectional E, E-CR, R or S-2 glass will be considered (Table 2.3). Because of the relative cheapness of E-glass fibre it is customary to use it together with a polyester or phenolic resin. The higher performance, and somewhat more expensive, R or S-2 glass, is frequently employed in conjunction with an epoxy matrix. Some basic properties of glass fibres are given in Table 5.1.

Observations

The strength refers to that of virgin filaments. Handling the filaments may reduce the strength noticeably. The specific heat measurements for S-2 glass refer to bulk material.

Table 5.1 Thermal and mechanical properties of glass fibres

property		E	R	S-2	E-CR
ρ	Mg/m^3	2.52–2.62	2.55	2.49	2.7–2.72
E_{lt}	GPa	73	86	86	72.5
σ_{lt}	MPa	3400	4400	4500	3300
ε	%	3.5–4.8	5.1	5.2–5.4	
ν_{lt}		0.22	0.215		
α	$°C^{-1}$	5×10^{-6}	4×10^{-6}	2.3×10^{-6}	5.9×10^{-6}
λ	W/m.K	1	1		
C_p	J/kg.K	740–840	840	800	

Notes: The first subscript l or t refers to directions longitudinal and transverse to the fibre. The second subscript, t, f or c, refers to tensile, flexural and compressive properties. For example, E_{tt} is the transverse tensile modulus.
Where neither l or t is shown, the property is isotropic in nature; for example α is the same for glass in all directions.
The symbols used here and elsewhere in this chapter are fully defined in Table 3.3.

Source

Vetrotex and Owens Corning Fiberglas trade literature.

■ 5.2 GLASS FIBRES

■ **Glass type**

Mechanical properties

Mechanical Properties

Some basic mechanical properties of unidirectional glass fibre composites are given in Table 5.2.

Materials

The first two materials are based on an unspecified epoxy resin, though the resin and size will be compatible. The second two composites are made with Stratyl 108, a polyester resin. The S-2 glass composite is an epoxy prepreg system. Manufacture was achieved by contact or prepreg moulding.

Observations

Some ILSS results are for a 50 v/o fibre loading. The longitudinal properties are clearly dominated by those of the fibre, while transverse tensile moduli are similar to those given later in the chapter for aramid and carbon fibre composites.

The product of the fibre modulus or strength and the fibre volume fraction should, according to the law of mixtures, and ignoring the contribution of the much less stiff and strong matrix, approximate to the composite modulus and strength. The ratio of the measured property to that calculated is known as the reinforcement efficiency. The value of this quantity is very high for the composite modulus irrespective of the type of fibre or matrix.

However for strength the efficiency is lower and is strongly influenced by the matrix type. The figures for E- and R-glass in an epoxy matrix are approximately 70%, but only 36% for E-glass in polyester, 46% for R-glass in polyester and 54% for S-2 glass in epoxy. The reduced reinforcement efficiency for the strength of glass fibre composites is due in part to damage sustained by the fibres when handled to make a composite. This effect may be worse for polyesters than for epoxies

Table 5.2 Basic mechanical properties of glass fibre composites

property		E-glass epoxy	R-glass epoxy	E-glass polyester (Stratyl 108)	R-glass polyester (Stratyl 108)	S-2 glass epoxy prepreg
V_f	v/o	60	60	42	42	71
ρ	Mg/m^3	2	2	1.8	1.8	
E_{lt}	GPa	37–47	53	31.6	37.5	47.6
σ_{lt}	MPa	1200–1600	1900	514	855	1732
σ_{lf}	MPa	1200	1500	622	784	1245
E_{tt}	GPa	9	13.6	7		
ν_{lt}		0.31	0.32	0.35		
ν_{tl}		0.09	0.09	0.11		
ILSS	MPa	72*	83*	53	54	83

* For 50 v/o fibre loading.

because the latter specimens may have been made with a prepreg which could minimize fibre damage. Some of the information in Table 5.2 is shown in histogram form in Figure 5.1.

Design implications

For a given volume loading of fibre, R-glass consistently gives higher longitudinal composite properties than E-glass, due to the intrinsically higher properties of R glass (Table 5.1). This must be balanced against the greater cost of the R glass reinforcement.

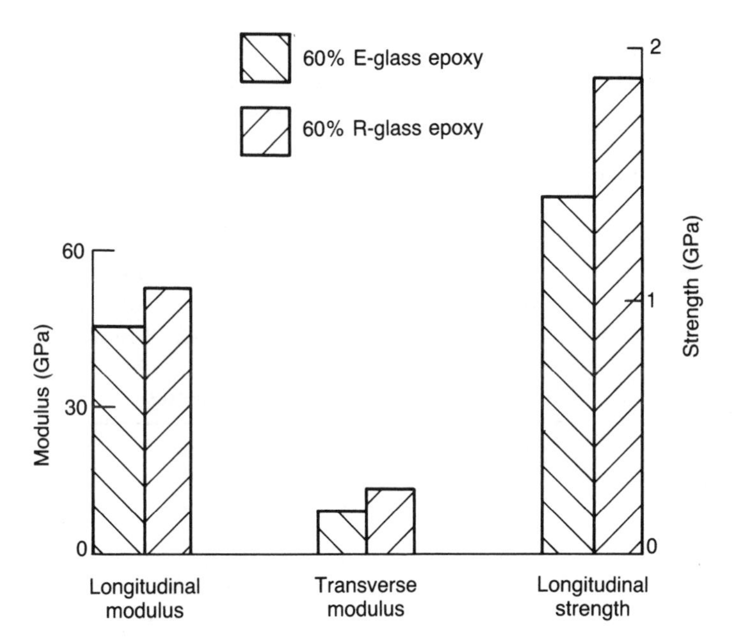

Figure 5.1 Longitudinal and transverse modulus and longitudinal strength data for E- and R-glass epoxy composites.

Source

Vetrotex and Owens Corning Fiberglas trade literature.

■ 5.2 GLASS FIBRES

■ **Fibre orientation**

Ply lay-up

The effects of fibre orientation

The anisotropy, or directional nature, of unidirectional fibre composites is mentioned in Chapter 1. To improve the modulus and strength for intermediate angles (i.e. between 0° and 90°) woven fabrics or multiple constructions are used. The latter is made up of a series of unidirectional plies laid up so that there is an angle, say, 10°, 20°, 30° or 45°, etc., between successive plies. To avoid the laminate distorting it is necessary to balance the construction about the centre plane – i.e. to have as many $-\theta$ plies as $+\theta$ ones. A typical balanced laminate is $[0 \pm 45\ 0]_S$. Because there are now some plies in intermediate directions the modulus and strength of the laminate in these directions is increased. The exact values for thermoelastic properties can be calculated from classical laminate theory, see Jones (1975). It is more difficult to calculate the effect on strength because of interaction between failure modes and individual plies, etc.

Some experimental results for filament wound E-glass polyester plates showing the effects of different ply constructions are given in Table 5.3.

Further information on the effect of ply construction on the initial modulus, ultimate tensile strength and ultimate tensile strain, of a $[\pm\theta]_s$ glass fibre composite is shown in Figures 5.2, 5.3 and 5.4.

Table 5.3 Effect of lay-up on glass fibre composite properties

lay-up number of plies		0° 8	±10° 8	±60° 8	$[0_2 \pm 45° \ 0]_s$ 10
V_f	v/o	56.6	51.1	58.1	55
E_{lt}	GPa	47.2	40.9	16.5	29.7
σ_{lt}	MPa	862	546	55	655
ε_{lt}	%	2.3	1.5	0.5	2.7
E_{tt}	GPa	14.7			
σ_{tt}	MPa	43			
ε_{tt}	%	0.3			
υ_{lt}		0.31			

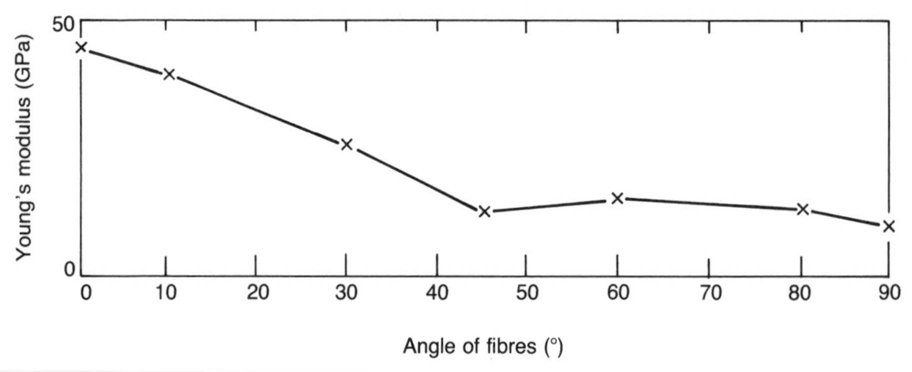

Figure 5.2 Off-axis modulus data for a glass fibre polyester composite. (Reproduced with permission of Anderson, S.I. and Lilholt, H., Risø National Laboratory, Denmark.)

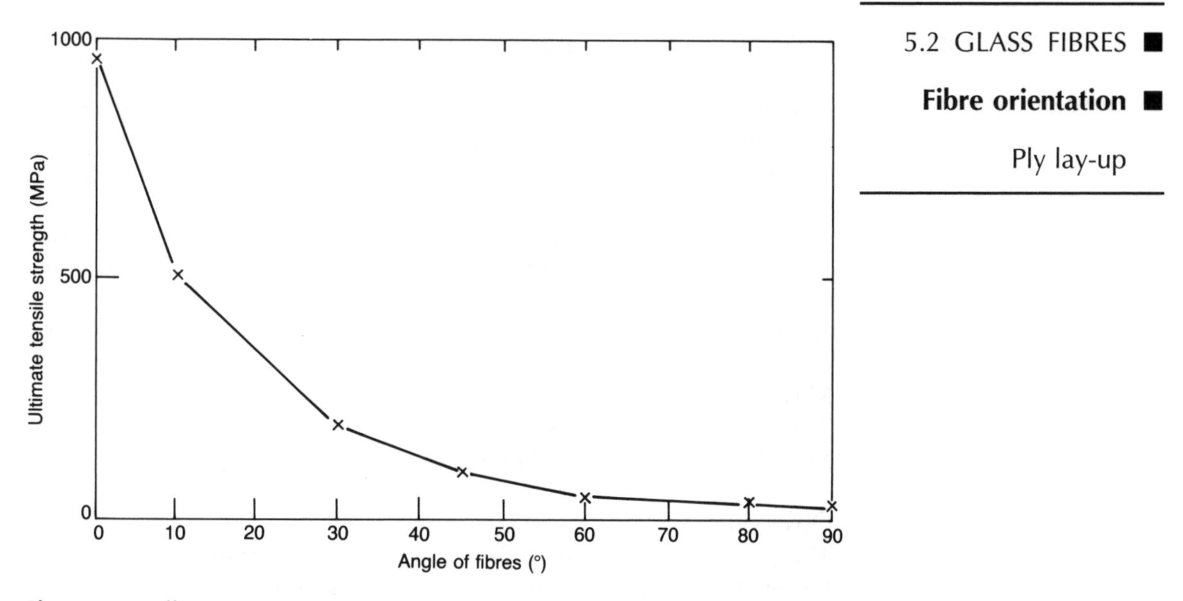

Figure 5.3 Off-axis strength data for a glass fibre polyester composite. (Reproduced with permission of Anderson, S.I. and Lilholt, H., Risø National Laboratory, Denmark.)

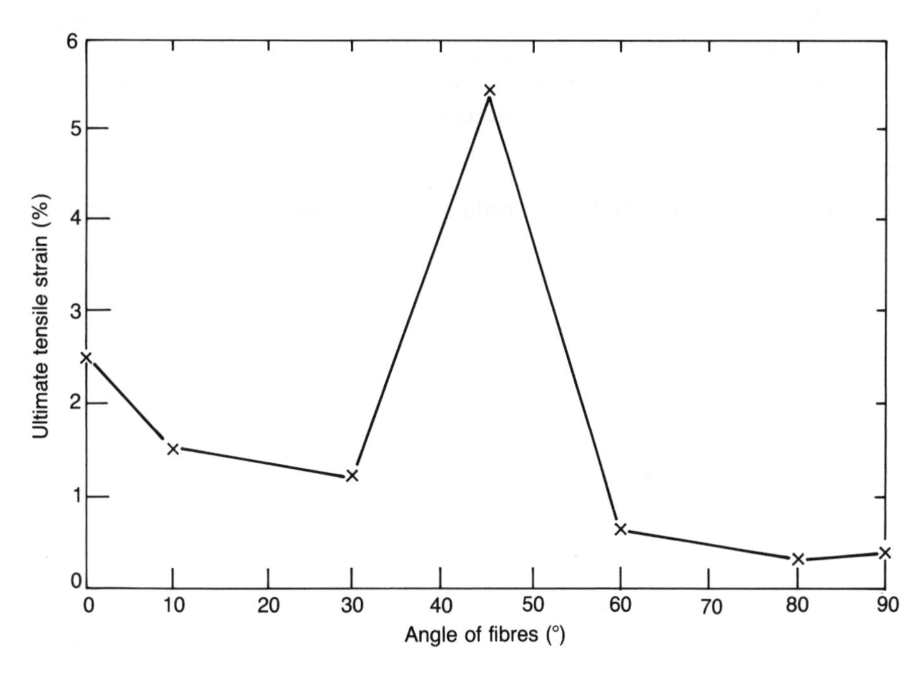

Figure 5.4 Off-axis strain data for a glass fibre polyester composite. (Reproduced with permission of Anderson, S.I. and Lilholt, H., Risø National Laboratory, Denmark.)

■ 5.2 GLASS FIBRES

■ **Fibre orientation**

Ply lay-up

The direction of measurement is along the 0° axis in all cases. The tensile modulus falls steadily until the ply construction is ±45° and then remains more or less constant. The ultimate tensile strength falls more dramatically to about 10% of its initial value for a unidirectional composite, for a ±45° ply construction. The failure strain peaks markedly for a ±45° construction. Results for carbon fibre composites are similar in pattern for strength, but the modulus shows a much steeper initial decrease than that for the glass fibre composite and the strain maximum, for the carbon fibre composite is much less, being barely greater than the 0° value.

Materials

The materials are E-glass rovings in Hoechst UP333 iso-polyester resin. Manufacture was achieved by filament winding plates.

Observations

The $[0_2 \pm 45\ 0]_S$ laminate data in Table 5.3 illustrate the advantage, for modulus and strength, of combining unidirectional and off-axis plies. The data show the susceptibility of strength in particular to an off-axis ply construction, though with the bonus of much improved strain-to-failure at ±45°.

Design implications

It is essential when working with composites, particularly unidirectional systems, to identify load paths accurately and allow for off-axis stresses, which may be regarded as secondary in nature, by placing reinforcement in those directions as well. If this is not done failure may occur since most composites lack the isotropic nature of for example many metals which allows them to accommodate secondary stresses.

Source

S.I. Anderson and H. Lilholt at the Risø National Laboratory, Denmark.

The influence of temperature

The effects of long-term exposure at temperature are given in Table 5.4.

Materials

The composites comprise continuous E-glass rovings and standard laminating systems.

Observations

The values all represent the percentage retention of a property. Glass fibre-reinforced composites can retain a high proportion of their initial longitudinal strength and modulus after prolonged exposure at temperature. Note that exposure times at temperature are different due to availability of data. It is interesting to compare the data here with that in Table 3.2. The latter is general information for polymers continuously exposed to an elevated temperature. The information in Table 5.4 is for composite properties. Reasonable instantaneous performance above the normal working temperature of the matrix is not unexpected. BS 5480 recommends that polymers are not used within 20°C or less of their HDT. However the HDT depends on the grade of the polymer, for a thermoset the cure conditions, and the presence of any reinforcement.

Some of the information in Table 5.4 is shown in histogram form in Figure 5.5.

Design implications

The data show that it is possible to use glass fibre composites at elevated temperatures. As the data are 10 to 15 years old they must now be regarded as indicative of high temperature capability.

Table 5.4 Percentage stiffness and strength retention of glass fibre composites at elevated temperature

property		epoxy	phenolic	bismaleimide	PPS	PEEK
instantaneous test at	°C	200	300	175	250	300
E_{lt}	%	90	30	97	21	61
σ_{lt}	%	60		80	26	
after exposure at	°C	200	250	250	246	
time	hours	10800	100	2000	10000	
E_{lt}	%			23		
σ_{lt}	%	69	59	77	50	
test temperature	°C	200	250	200	20	

■ 5.2 GLASS FIBRES

■ **Resin type**

Temperature

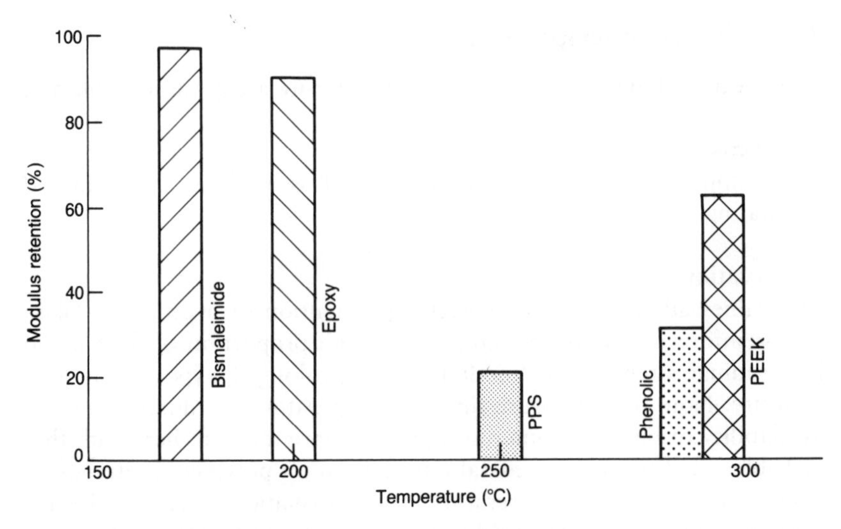

Figure 5.5 Percentage instantaneous modulus retention of glass fibre composites with various resins.

Source

Hancox, N.L. and Phillips, D.C. (1985) *Proc. 2nd Materials Engineering Conference*, London, 139–44.

Thermoplastic matrix systems

Short glass fibres are frequently compounded with thermoplastic matrices. Here some information relating to continuous, unidirectional, fibre materials is given (Table 5.5).

Materials

The thermoplastics, PPS (polyphenylene sulphide) and PAS (polyaryl sulphone), are both manufactured by Phillips Petroleum. The reinforcement is an unspecified E-glass. Manufacture was achieved by laying-up preimpregnated tape.

Observations

The fibre loading of the tape used to make the PAS specimens was 53 v/o. Judging from the modulus results some extra compression has taken place during fabrication to increase this figure slightly. The reinforcement efficiency for the modulus is excellent, though for the strength the figures are not so high especially for the PPS material. This is probably indicative of damage to the glass fibre in handling during fabrication and the fact that the surface treatment of the fibre may not have been optimized for these types of matrix. Transverse properties are similar to those for other glass- and fibre-reinforced materials and the compressive performance good.

Design implications

Regarding mechanical performance thermoplastic matrix composites can be as useful as thermoset matrix systems.

Table 5.5 Properties of glass fibre-reinforced thermoplastics

property		PPS/E-glass	PAS/E-glass (Phillips Petroleum)
V_f	v/o	70	
ρ	Mg/m^3	2	
E_{lt}	GPa	49.7	43.5
σ_{lt}	MPa	911	1118
E_{tt}	GPa		13.8
σ_{tt}	MPa		46
E_{lf}	GPa	49	41.4
σ_{lf}	MPa	759	1111
E_{lc}	GPa	44.2	42.6
σ_{lc}	MPa	1159	1201
ILSS	MPa		63

Source

Phillips Petroleum trade literature.

■ 5.3 ARAMID FIBRES

■ **Properties**

5.3 ARAMID FIBRES

There are several sources of aramid fibre, principally Kevlar, made by du Pont, and Twaron and Technora, made by Akzo and Teijin respectively. Chemically the first two are similar but differ somewhat from the last. Kevlar fibres, and presumably the other types, have a microfibrilar structure which may account for the relatively poor performance of aramid composites under compression and the high moisture regain of the fibre. Aramid fibres are of interest to the user of composites because of their very low density, good range of moduli, high strength and strain of failure, thoughness and handleability. They are more expensive that glass fibres and have a lower temperature capability and poor compression strength. The fibres are available in various physical forms including woven fabrics and with different finishes to aid bonding to specific matrices. Some fibre properties are listed in Table 5.6.

Observations

For a given type of fibre the properties quoted in data sheets and the literature do vary somewhat. This is due to several factors including the continual development and improvement of the reinforcement over a period of time and the effects of test technique on the property measured.

Table 5.6 Properties of aramid fibres

property		Kevlar 29	Kevlar H_p	Kevlar 49	Kevlar H_m	Twaron	Twaron HM	Technora
ρ	Mg/m^3	1.44	1.44	1.45	1.47	1.44	1.45	1.39
E_{lt}	GPa	58	100*	117*	160†	80	115	73
σ_{lt}	MPa	2760	2960	2760	2400	2800	2800	3400
σ_{lt}§	MPa	3620	3590	3620	3450	3150	3150	
ν_{lt}				0.36‡				
ε	%	3.6	3.1	1.9	1.4–1.5	3.3	2.0	4.6
α_l	$°C^{-1}$			-5.2×10^{-6}‡		-3.5×10^{-6}	-3.5×10^{-6}	-6×10^{-6}
α_t	$°C^{-1}$			41.4×10^{-6}‡				
λ	W/m.K					0.04	0.04	
C_p	J/kg.K					1420	1420	1090

* 1% secant modulus.
† 0.5% strain modulus.
‡ The property was determined from measurements on a composite.
§ The strength was measured on an epoxy impregnated strand as per ASTM D2343.

Source

du Pont, Akzo and Teijin trade data.

Composite properties

The properties of some unidirectional aramid fibre epoxy resin composites are given in Table 5.7.

Materials

- resins – (1) Shell Epon 828 epoxy, cured with MNA and BDMA, (2) Fiberite 934, an epoxy resin containing tetrafunctional epoxy, (3) Bakelite epoxy, (4) Ciba Geigy LY556 epoxy with HY917 hardener and DY070 accelerator;
- fibres – Kevlar fibres are made by du Pont. H_m is a development product of 49 with a higher modulus. Kevlar H_m is known as Kevlar 149 in USA. Twaron is an aramid fibre made by Akzo.

Observations

These fibres are representative of the high performance aramid products currently available. The anisotropy of unidirectional composites is apparent as is the poor compression strength of aramid composites. The composite densities are extremely low.

Table 5.7 Properties of undirectional aramid composites

property		Kevlar 49 (1)	Kevlar H_m (1)	Kevlar H_m (2)	Kevlar 49 (2)	Twaron HM (3)	Twaron HM (4)
V_f	v/o	60	60	58	58	64–68	64–68
ρ	Mg/m^3	1.36	1.37	1.35	1.37	1.4	1.40
E_{lt}	GPa	79	107	106	72	66.7	
σ_{lt}	MPa	1500	1450	1400	1520	1291	
ε_{lt}	%	1.71	1.33	1.3	1.6		
E_{tt}	GPa	4.1					
σ_{tt}	MPa	27					
ε_{tt}	%	0.85					
E_{lf}	GPa	67	79			45.2	
σ_{lf}	MPa	655	634			655	
ε_{lf}	%						
E_{lc}	GPa	66	73	66	64	63.4	
σ_{lc}	MPa	234	193	262	280	265	253
ε_{lc}	%	0.58					
E_{tc}	GPa	5.2					
σ_{tc}	MPa	93					
ε_{tc}	%	2.83					
υ_{lt}		0.43		0.34	0.41		
υ_{tl}		0.31					
G_{lt}	GPa	1.5					
G_{tl}	GPa	1.5					
τ_{lt}	MPa	47		49	47		
τ_{tl}	MPa	27					
ILSS	MPa	48–69	57	38	50	54	45

Notes: The Twaron data were taken from measurements made according to DIN 29971.
Kevlar H_m is a development of 49 with a higher modulus (known as Kevlar 149 in USA).

Design implications

Care must be taken with compressive loading. Other aspects are discussed subsequently.

Sources

Pindera, M.J., Gurdal, Z., Herakovich, C.T. *et al.* (1989) *Journal of Reinforced Plastic Composites*, **8**, 410–19.

Reedy, E.R. (1988) *Journal of Composite Materials*, **22**, 955–65.

du Pont and Akzo data sheets, and especially R. Pinzelli of du Pont, Geneva.

The effect of resin type

To illustrate the effect of the type of epoxy resin on composite properties consider the entries in the first, fourth and fifth columns of Table 5.7, which are for an anhydride cured epoxy, Epon 828, a largely tetrafunctional aerospace epoxy, Fiberite 934 and a Bakelite epoxy respectively, but very similar types of reinforcement. Some properties are given in Table 5.8.

Materials

- resins – Shell Epon 828 epoxy, cured with MNA and BDMA, Fiberite 934, an epoxy resin containing tetrafunctional epoxy, Bakelite epoxy;
- fibres – Kevlar 49 is made by du Pont. Twaron HM is an aramid fibre made by Akzo.

Observations

The Epon and Fiberite matrices appear to give a better translation of most fibre properties into composite properties. This may be due to better fibre wet-out or fibre/resin bonding.

Design implications

The correct choice of resin, and hence fibre/resin wet-out, is important in attaining the best mechanical properties. The designer should remember this when specifying materials and may have to do some preliminary experimental work to select the right system.

Table 5.8 Effect of resin type on properties of aramid composites

property		Epon 828	Fiberite 934	Bakelite epoxy
V_f	v/o	60	58	64–68
E_{lt}	GPa	79	72	66.7
σ_{lt}	MPa	1500	1520	1291
ε_{lt}	%	1.71	1.6	
E_{lc}	GPa	66	64	63.4
σ_{lc}	MPa	234	280	265
ILSS	MPa	48–69	50	54

Sources

As per Table 5.7.

■ 5.3 ARAMID FIBRES

■ **Epoxy resin**

Fibre modulus

The effect of fibre modulus

A similar analysis to that given in the previous section can be given for the effect of modulus on composite properties. Consider the first two columns of Table 5.7, which are for Kevlar 49 and Kevlar H_m fibre respectively in the same type of epoxy resin matrix.

Materials

The epoxy resin used was Shell's Epon 828/MNA/BDMA. Reinforcement was with aramid fibres: Kevlar 49 and Kevlar H_m (du Pont).

Observations

The modulus of the composite increases and the strength decreases approximately in proportion to that of the fibre as would be expected.

Design implications

The designer should always attempt to take an 'all round' view of the problem and beware of concentrating on one property at the expense of the others.

Table 5.9 Effect of fibre modulus on composite properties

property		Kevlar 49	Kevlar H_m
V_f	v/o	60	60
E_{lt}	GPa	79	107
σ_{lt}	MPa	1500	1450
ε_{lt}	%	1.71	1.33
E_{lc}	GPa	66	73
σ_{lc}	MPa	234	193
ε_{lc}	%	0.58	

Source

As per Table 5.7.

The effect of fibre volume loading

Most practical composites contain, overall, 55 to 65 v/o of fibre, though the fibre will often be laid up in more than one direction. To make composites with much less fibre is difficult as the fibres tend to lose their orientation, bunch, etc., while to get a much higher volume loading leads to the fibres being crushed together, high stress concentration, etc. The rule of mixtures for modulus and strength indicates, to a first approximation, a linear relationship between fibre and composite longitudinal modulus and strength.

Table 5.10 and Figure 5.6 give some strength data for Kevlar 49 fibres in an epoxy resin which shows the experimental and calculated effects of a wide variation in V_f. Individual entries are the average of five to ten readings.

Materials

The epoxy resin used was Ciba Geigy's MY750 with HT972 hardener. Reinforcement was du Pont's Kevlar 49 aramid fibre.

Observations

It is noticeable that the measured strength as a percentage of the calculated value peaks in the vicinity of 50 v/o fibre loading. Below this value of V_f it was suggested that non-homogeneity and misorientation reduce the measured tensile strength below that expected while above $V_f = 50$ v/o lack of matrix penetration between fibres and fibre damage due to squeezing reduce expected properties.

A similar pattern would be expected for other types of fibre reinforcements. Data are not available for transverse and shear properties but because of the low values of these parameters compared with the fibre-controlled properties, the effects are unlikely to be as great.

Design implications

It is tempting, on the basis of the law of mixtures, to seek to improve modulus or strength in a composite by increasing the volume fraction of

Table 5.10 Effect of fibre fraction on tensile strength of aramid composites

V_f (v/o)	σ_{lt} (MPa)	σ_{lt} measured / σ_{lt} calculated
26	580 ± 50	0.71
38	860 ± 30	0.76
40	930 ± 70	0.83
46	1120 ± 30	0.81
52	1376 ± 50	0.9
59	1390 ± 40	0.79
64	1370 ± 90	0.72
68	1480 ± 120	0.73
73	1500 ± 60	0.69

■ 5.3 ARAMID FIBRES

■ **Epoxy resin**

Fibre volume

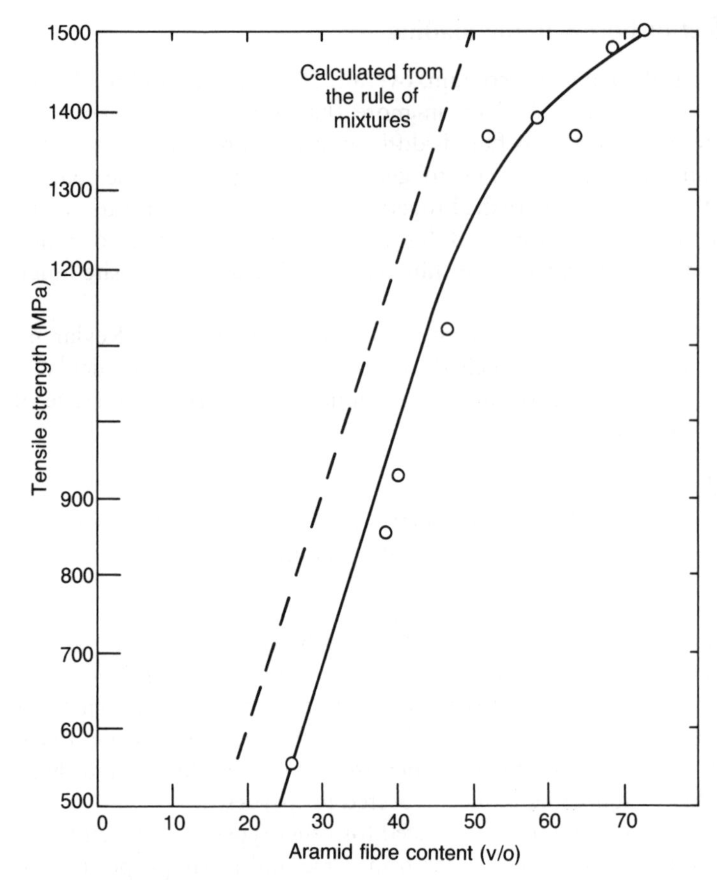

Figure 5.6 Tensile strength of unidirectional Kevlar 49 composite against fibre fraction. (Redrawn from Mittleman, A. and Roman, I. *Composites, 21*, 63–9; published by Butterworth-Heinemann, 1990.)

fibre. For an aramid fibre, and probably for other types, this begins to show diminishing returns above approximately 60 v/o. If very high tensile properties are required it may be better to use another type of fibre with a higher modulus or strength or rethink the design rather than increase the fibre loading too much.

Source

Mittleman, A. and Roman, I. (1990) *Composites*, **21**, 63–9.

Shear properties and the influence of test methods

Because of their heterogeneous nature and, sometimes, extreme anisotropy it has been necessary to develop special test methods for composite materials (Mayer, 1993). This work is still underway, as is the debate on which methods are preferred. The effects of test method, shape and fibre orientation on the measured shear properties of Kevlar 49 fibre composites are given in Table 5.11.

Materials

The results are for Kevlar 49 fibre composites prepared with an epoxy resin, Dow Chemicals DER 332, cured with an amine. The fibre volume loading was 65 v/o.

Observations

Four of the five methods give similar results. For the Kevlar material the shear stress/strain characteristic was linear up to 0.5% strain in each case.

Design implications

Aramid fibre composites have a relatively low shear strength and modulus. The influence of test method on the data produced must always be borne in mind when selecting design data.

Table 5.11 Effect of test method on shear properties

test method	shear strength (MPa)	shear strain at failure (%)	shear modulus at 0.5% strain (GPa)
torsional shear			
90° wound thin walled tube	31.3 ± 2.0	2.0 ± 0.2	1.74 ± 0.4
0° wound composite rod	31.3 ± 2.8	1.82 ± 0.2	1.96 ± 0.09
tensile stressing			
±45° symmetrical laminate	29.4 ± 0.7	1.73 ± 0.5	1.92 ± 0.11
10° off-axis specimen	19.4 ± 1.1	0.97 ± 0.08	2.08 ± 0.1
ILSS	36.8 ± 0.9		

Source

Chiao, C.C., Moore, R.L. and Chiao, T.T. (1977) *Composites*, July 1977, 161–9, by permission of University of California, Lawrence Livermore National Laboratory and the United States Department of Energy under whose auspices the work was performed, and Butterworth Heinemann Ltd., publishers. Note that the US Government retains the right to non-exclusive royalty-free in and to any copyright covering this material.

■ 5.3 ARAMID FIBRES

■ **Temperature**

Tensile strength

The influence of temperature on the properties of aramid fibres

Materials

Kevlar 49 fibre is made by duPont and Technora by Teijin.

Observations

The Kevlar samples were tested in the form of epoxy resin roving strands; the Technora as filament yarns (Table 5.12).

Design implications

At temperatures of up to 180°C the strength retention is good. The use of a suitable resin matrix would protect the fibre from oxidation and probably improve the life-time at a given temperature.

Table 5.12 Effect of temperature on the properties of aramid fibres

fibre	temperature (°C)	exposure time (hours)	tensile strength retained (%)
Kevlar 49 – as epoxy resin roving strands (Du Pont)	Up to 150	168	100
	200	168	88
	250	168	49
Technora – as filament yarns (Teijin)	180	1000	87
	200	1000	70
	250	1000	33
	300	100	38
	350	10	34

Sources

Nose, K. and Okamoto, T. (1990) *Proc. Seminar at PIRA International*, Leatherhead, December 1990.

du Pont trade data.

5.4 CARBON FIBRES

5.4 CARBON FIBRES ■

Mechanical properties ■

Unlike other reinforcements there are over 100 types of carbon fibre available. Carbon fibres have densities between those of aramids and glass fibres. The fibres are strong, can have a very high modulus, good fatigue resistance (in composite form), may have good electrical and thermal conductivity and very low thermal expansion in the direction of the long fibre axis. They may however be rather brittle, if unprotected oxidize above 300–400°C, and are more expensive than the other types of reinforcement.

Variables, which should be considered when fabricating with carbon fibres, are the number of filaments per tow – 1, 3, 6, 12 $\times$ 10^3 are common values, but greater numbers are available for some lower strength carbon fibres – and the protective size coupling layer on the surface of the fibre. Prior to being coated most carbon fibres are surface treated to enhance bonding, and it is desirable to have a size which is compatible with the resin system being used. As with glass fibres details of coatings are closely guarded and the manufacturer should be consulted about a suitable system for use with a specific resin matrix. Generally the higher the fibre modulus the better aligned are the crystallites in the structure of the fibre, the lower the strain to failure and the more expensive the material. The cost rises with processing temperature.

Specifying the properties of carbon fibre composites is difficult because of the very large number of fibre types and the continual development of new varieties. Much of the information generated prior to the early 1980s may no longer be relevant due to a reduction in fibre variability and improvements in manufacturing techniques, better fibre/resin bonding and because the fibre may no longer be available. It is tempting, given the amount of information available, to use simple models (e.g. the rule of mixtures) to predict longitudinal composite strength and modulus. Though this will give an estimated, upper, value of the property, the translation from fibre to composite performance is not always simple. Factors which may reduce composite properties include fibre misalignment during composite manufacture, too high a fibre volume loading leading to excessive stress concentration between fibres, incompatibility between the fibre coating and resin matrix, thermally induced stresses caused by curing at a high temperature, etc. Thus a more reliable way to assess composite properties is to make use of information generated on well-made test plaques. Unfortunately many non-fibre dominated properties are infrequently measured and in these cases it may be necessary to refer to results from older studies.

■ 5.4 CARBON FIBRES

■ **Mechanical properties**

Properties of carbon fibres

Characteristic properties of a range of carbon fibres are listed in Table 5.13.

Materials

The carbon fibre reinforcements illustrated are: AS4, IM7 (Hercules); XA, HM400 (Courtaulds); T300, T800H, M46J, M60J (Toray); IM600 (Akzo); P55S, P100S, P120S (Amoco).

Table 5.13 Properties of carbon fibres

property	high strength			intermediate modulus		
	AS4	XA	T300	T800H	IM600	IM7
E_{lt} GPa	224	230	235	294	294	303
σ_{lt} GPa	4.0	3.9	3.53	5.49	5.64	5.3
ε_{lt} %	1.6	1.7	1.5	1.9	1.9	1.77

property	high modulus			ultra high modulus		
	P55S	HM400	M46J	M60J	P100S	P120S
E_{lt} GPa	380	405	436	588	724	827
σ_{lt} GPa	1.9	3.1	4.21	3.92	2.2	2.2
ε_{lt} %	0.5	0.8	1.0	0.7	0.31	0.27

Observations

There is a very wide range of properties which is also reflected in the cost. In general, the higher the modulus, the higher the price.

Source

Trade data.

Carbon fibre composites based on thermosetting matrices

Table 5.14 summarises some properties of primarily, epoxy resin matrix composites.

Materials

- resins – (1) Ciba-Geigy 913, 120°C cure epoxy, (2) Courtauld's toughened epoxy, 175°C cure, (3) Hercules epoxy, high temperature performance, (4) Hercules epoxy, high temperature performance, T_g (dry) = 190°C, T_g (wet) = 139°C, (5) Hercules epoxy, high temperature performance, T_g (dry) = 215°C, T_g (wet) = 154°C, (6) US Polymerics bismaleimide resin, (7) Toray epoxy 2500, 121°C cure, (8) Toray epoxy 3631, 177°C cure, semi-toughened;
- fibres – XA and IM 43 carbon fibres are made by Courtaulds, the AS4 fibre by Hercules, and the remainder, the T and M series, by Toray. Note: Courtaulds have ceased producing carbon fibres but similar grades are available from other manufacturers.

Manufacture

Specimens would be prepared from a prepreg or by a wet lay-up method. All specimens are unidirectional.

Observations

The carbon fibres used are from the separate sources and cover all the grades listed in Table 3.1. Measurements were made at room temperature and all the matrices except one are epoxies. The use of bismaleimide resin makes little difference to properties apart from the relatively low transverse strain-to-failure. The large gaps in information especially for transverse, compressive and shear properties are clear. The anisotropy of unidirectional materials can be seen by comparing longitudinal and transverse tensile or shear moduli and strengths.

Design implications

Longitudinal tensile properties are excellent, longitudinal compressive properties very good, but anisotropy needs to be watched.

Sources

Ramkumar, R.L., Grimes, G.C. and Kong, S.J. (1986) *Special Technical Publication 893*, ASTM, 48–63.

Courtaulds, Hercules and Toray trade data.

Table 5.14 Properties of unidirectional carbon fibre composites

property		XA[1]	IM-43[2]	AS4[3]	AS4[4]	AS4[5]	T300[6]	T300[7]	T300[8]	T800[7]	T800H[8]	M46J[7]	M46J[8]	M60J[7]	M60J[8]
V_f	v/o	60	60	60	60	62	62	60	60	60	60	60	60	60	60
ρ	Mg/m^3	1.57	1.54	1.56	1.56	1.66	1.57	1.54	1.54	1.57	1.57	1.59	1.59	1.65	1.65
E_{lt}	GPa	138	175	148	148	148	133	125–140	125	150–170	160	245–260	260	310	330
σ_{lt}	MPa	2000	2924	2137	2137	2137	1510	1760–1800	1760	2840–2900	2840	2200–2500	1960	1960	1760
ε_{lt}	%						1.13	1.3	1.3	1.6	1.55	0.9	0.7	0.6	0.5
E_{tt}	GPa		10				9	8.8–9.0	7.8	8.8–9.0	7.8	6.9–7.5	6.9	5.9	5.9
σ_{tt}	MPa		71				34	80	80	65–70	80	45	45	30	30
ε_{tt}	%						0.358	1.0	1.0	0.8–0.9	1.0	0.6		0.6	
E_{lf}	GPa	126		134	128	127		120		150		235			
σ_{lf}	MPa	180		1724	1793	1793		1700		1800		1450			
E_{lc}	GPa	126	172				130	125	125		145	230	245		320
σ_{lc}	MPa	1350	1600				1435	1370–1400	1570	1570–1600	1570	1030–1050	880	880	780
ε_{lc}	%						1.19		1.2						
ν_{lt}		0.25					0.31	0.34	0.34		0.34				
ν_{tl}		0.02					0.027								
G_{lt}	GPa								4.4		4.5		3.9		3.9
τ_{lt}	MPa								98		98		59		39
ILSS	MPa	100	112	127	127	120		100	110	100	100	90	80	70	70

The influence of resin type on properties

An idea of the effect of the resin matrix on the properties of composites made with the same grade of fibre, T300, and fibre volume loading, 60 v/o, is given in Table 5.15. The information has been taken from entries 7 and 8 of Table 5.14.

Materials

The resins used were Toray's epoxy 2500 with a 121°C cure and epoxy 3631 with a 177°C cure. Reinforcement was Toray's T300 carbon fibre.

Observations

There is little to choose between the two epoxies. The longitudinal tensile modulus is what would be expected from the rule of mixtures, section 1.3, though the strength is below the expected figure. The extreme anisotropy of unidirectional composites is noticeable and in both cases compressive strength is less than tensile strength.

Design implications

The composite properties are primarily determined by the reinforcement rather than the type of epoxy resin.

Table 5.15 Effect of resin type on carbon composite properties

property		epoxy 2500 121°C cure	epoxy 3631 177°C cure
E_{lt}	GPa	125–140	125
σ_{lt}	MPa	1760–1800	1760
E_{tt}	GPa	8.8–9	7.8
σ_{tt}	MPa	80	80
σ_{lc}	MPa	1370–1400	1570
ILSS	MPa	100	110

Source

As per Table 5.14.

■ 5.4 CARBON FIBRES

■ **Fibre type**

Mechanical properties

The influence of fibre type on properties

The influence of the fibre type on the properties of four 60 v/o composites is listed in Table 5.16. The data are for entries 7, 9, 11, 13 of Table 5.14. These are all for composites made using 2500 epoxy.

Materials

The resin was Toray's epoxy 2500 with a 121°C cure. The reinforcements were Toray's T300, T800H, M46J and M60J carbon fibres.

Observations

In terms of reinforcement efficiency using the higher property value quoted the modulus of the first three types of composite is excellent, while that of the fourth indicates an efficiency of 88%. The efficiencies for strength are 85, 88, 100 and 83%, respectively, indicating excellent fibre bonding. Transverse properties show a more or less steady decline with increasing fibre modulus as does compression strength overall. The last two observations are probably due to the more ordered fibre structure which is associated with increasing fibre modulus.

Design implications

While it is possible to utilize virtually all the fibre modulus in a composite this is not necessarily the case for fibre strength. Furthermore as the fibre and hence composite modulus is increased transverse and compressive performance tends to decrease. This must be carefully allowed for in design.

Table 5.16 Effect of fibre type on composite properties

property		T300	T800	M46J	M60J
E_{lt}	GPa	125–140	150–170	245–260	310
σ_{lt}	MPa	1760–1800	2840–2900	2200–2500	1960
E_{tt}	GPa	8.8–9	8.8–9	6.9–7.5	5.9
σ_{tt}	MPa	80	65–70	45	30
σ_{lc}	MPa	1370–1400	1570–1600	1030–1050	880
ILSS	MPa	100	100	90	70

Note: For fibre properties, see Table 5.13.

Source

As per Table 5.14.

Properties of composites based on very high modulus fibres

Some mechanical properties of this type of unidirectional fibre composite are shown in Table 5.17.

Materials

The reinforcements are P75S, P100S and P120S carbon fibres made by Amoco. The resin is an epoxy ERL 1962.

Observations

Both modulus and strength efficiency are excellent, in some cases greater than 100%, indicating errors in the fibre or composite data. Transverse properties and compressive strength are low because of the highly ordered nature of the fibre structure. The transverse failure strain, which is strongly influenced by the fibre/matrix bond, is now greater than the longitudinal failure strain.

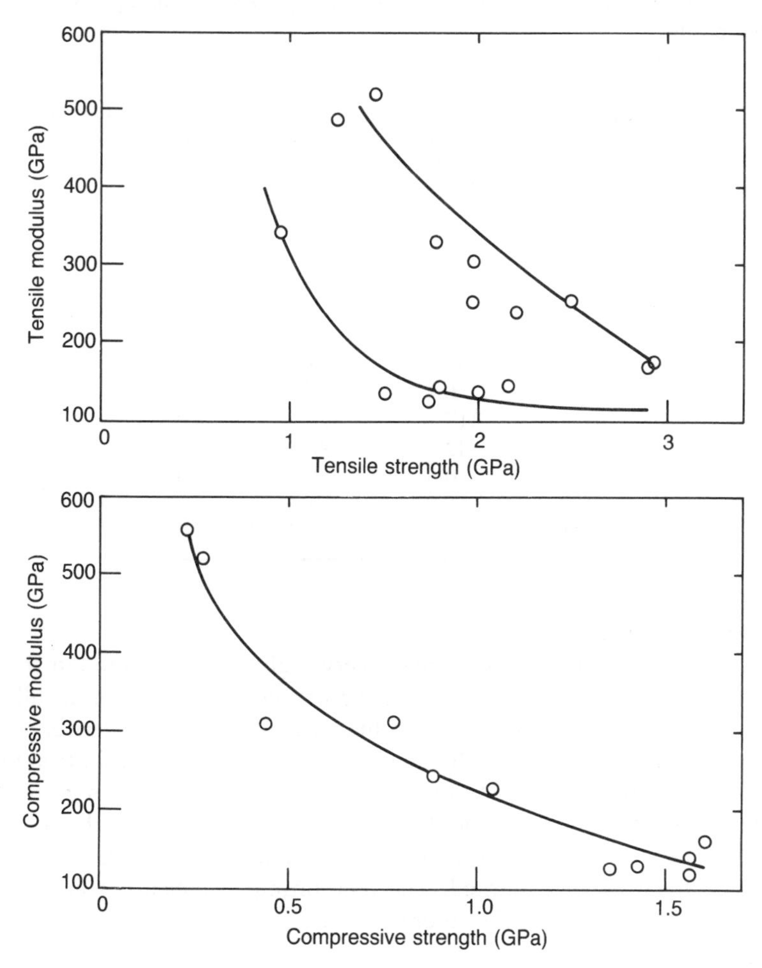

Figure 5.7 Relation between unidirectional composite tensile modulus and strength and compressive modulus and strength for carbon fibre composites.

■ 5.4 CARBON FIBRES

■ **Very high modulus fibres**

Mechanical properties

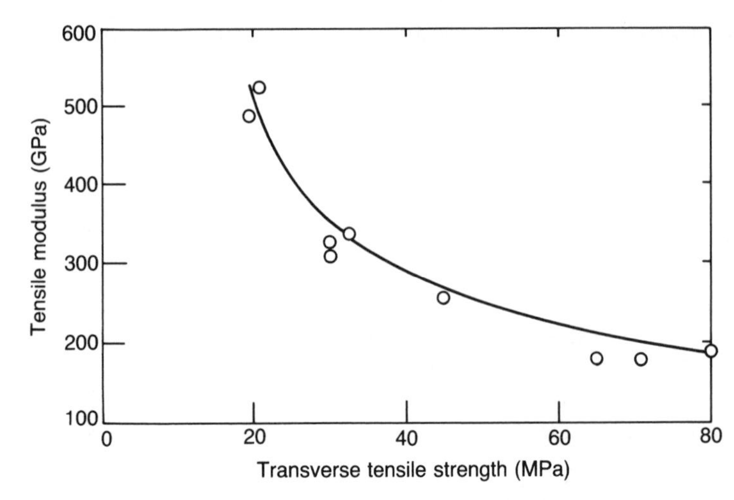

Figure 5.8 Relation between tensile modulus and transverse tensile strength of carbon fibre composites.

Table 5.17 Very high modulus carbon fibre composites

property		P75S	P100S	P120S
V_f	v/o	62	62	62
E_{lt}	GPa	338	497	524
σ_{lt}	MPa	966	1242	1442
ε_{lt}	%	0.28	0.24	0.27
E_{tt}	GPa	6.9	5.3	5.7
σ_{tt}	MPa	33	18	21
ε_{tt}	%	0.49	0.35	0.4
E_{lc}	GPa	317	524	559
σ_{lc}	MPa	442	290	270
ν_{lc}		0.3	0.31	0.29
G_{lt}	GPa	5.9	5.6	5.5
ILSS	MPa	55.2	34.5	27.6

Design implications

Although it is possible to achieve very high composite moduli the low transverse properties and compressive strength must be carefully allowed for in any design. In addition very high modulus fibres are very expensive and can prove difficult to handle because of their low strain-to-failure.

Some of the information in Tables 5.14 and 5.17 is shown graphically in Figures 5.7 and 5.8. The decrease in compressive strength and transverse tensile strength with increasing compressive and tensile moduli respectively is very marked. The relationship between tensile modulus and strength shows rather more scatter.

Source

Amoco trade literature.

Properties of bismaleimide resin composites

Bismaleimides or BMI resins are a family of thermosetting resins with an upper working temperature higher than that of an epoxy, i.e. 180–200°C compared with ~150°C. Some data, including an entry from Table 5.14, are shown in Table 5.18.

Materials

V378A is a US Polymerics bismaleimide resin. Compimide 795 800 and 65 FWR are resins manufactured by Shell, Germany. T300 is a Toray carbon fibre and Celion 6000 is made by BASF Structural Materials.

Observations

Generally the properties show little change with either fibre or resin type and for a similar fibre loading and type are similar to those noted for epoxies (Table 5.14).

Design implications

A resin can be chosen for its enhanced temperature performance without unduly influencing normal mechanical properties of the composite. Bismaleimides, though having a better upper working temperature than epoxies, are rather more expensive than the former.

Table 5.18 Properties of carbon fibre bismaleimide composites

property	resin fibre	V378A T300	65 FWR T300	Compimide 795 Celion 6000	Compimide 795 T300	Compimide 800 Celion 6000
V_f	v/o	62	60	60	60	
E_{lt}	GPa	133		119		
σ_{lt}	MPa	1510	1500	1450		
ε_{lt}	%	1.13				
E_{lf}	GPa		123	117	114	119
σ_{lf}	MPa		1800	1850	1920	1950
ILSS	MPa		110	96	120	99
E_{tf}	GPa		8.3	7.4	8.7	8.5
σ_{tf}	MPa		105	52	85	62
E_{lc}	GPa	130		128		
σ_{lc}	MPa	1435		1400		

Sources

Toray and Shell trade data.

■ 5.4 CARBON FIBRES

■ **Epoxy resin**

Transverse properties

The influence of the polymer matrix

The role of the polymer matrix can be complicated. Longitudinal composite properties are largely dependent upon those of the fibre while transverse and shear performance are much more dependent upon the matrix and fibre/matrix bond. In addition the matrix may influence longitudinal properties because of differences in stress transfer into and among fibres with different matrices and the ease with which the matrix penetrates fibre tows, which in turn may influence the formation of voids. Finally, the matrix largely determines the elevated temperature behaviour of the composite. The effects of matrix type on transverse properties are illustrated in Table 5.19.

Materials

3205 is a Hercules high temperature epoxy, based on a tetra-functional system cured with an amine. The other four epoxy resins are made by Hexcel. F155NR is cross-linked, rubber-modified, F155 is cross-linked, with no rubber, F185NR is cross-linked with a soft backbone and rubber-modified, while F185 is cross-linked with a soft backbone and contains no rubber. The F series resins will be tougher than the tetra-functional system. The fibre volume fraction was approximately constant, and the reinforcement was Hercules AS4 carbon fibre in each composite. All measurements were made at 23°C.

Observations

The best strength and modulus are achieved using the high performance and more brittle epoxy. The addition of a rubber as a toughening agent and the presence of a soft backbone between cross-links may increase the strain-to-failure but reduces other properties. It is probable that fibre/matrix bonding is important in determining the properties listed as well as the strain and fracture properties of the resin.

Design implications

It is sometimes assumed that a less brittle resin matrix will improve transverse composite performance but this is not always so. Great care must be taken in design in allowing for the effects of transverse (and shear) loads.

Table 5.19 Transverse properties of carbon composites measured at 23°C

property	3205	F155NR	F155	F185NR	F185
E_{tt} GPa	11.73	7.58	8.48	7.93	7.52
σ_{tt} MPa	66.2	53.6	62.8	43.2	49.8
ε_{tt} %	0.632	0.699	0.58	0.538	0.858

Source

Jordan, W.M., Bradley, W.L. and Moulton, R.J. (1989) *Journal of Composite Materials*, **23**, 923–43.

The influence of test methods on compressive properties

Because of their heterogeneous nature and, sometimes, extreme anisotropy, it has been necessary to develop special test methods for composite materials. Details are given by Mayer (1993). This work is still underway, as is the debate on which methods are preferred. Some details for compression measurements are given in Table 5.20.

Materials

The resin used was a Hercules epoxy with a high temperature performance; reinforcement was carbon fibre. Thirty-two plies of prepreg were used to make the specimens so that the fibre volume loading was approximately constant.

Observations

The coefficient of variability lay between 3.4 and 7.8%. The IITRI and Celanese compression testing jigs are standard designs that have been used for some time. Further work shows that the type of end tab material (steel or glass fibre composite), the tapering of the grip ends and the extent to which the tab was gripped all influenced the compressive strength of the carbon fibre composite. Full gripping and steel end tabs gave the highest results.

Design implications

The spread in modulus is 7%, comparable with the coefficient of variation noted, but the spread in strength is 19%, well above the experimental variation. Clearly, test methods must be carefully chosen and specified and it may be necessary to determine material properties in the context of a given design until the effects of test methods on properties is fully understood in theory and in practice.

Table 5.20 Effect of test method on the longitudinal compressive properties of carbon fibre composites

property	IITRI*	Wyoming modified Celanese	Wyoming end loaded, side supported
E_{lc} GPa	124	126	133
σ_{lc} MPa	1390	1240	1170

* Illinois Institute of Technology Research Institute.

Source

Berg, J.S. and Adams, D.F. (1989) *Journal of Composites Technology and Research*, **11**, 41–6.

■ 5.4 CARBON FIBRES

■ **Temperature effects**

Resin type

Temperature effects

Carbon fibres, if unprotected, will begin to oxidize at 300–350°C. The performance of fibre composites below these temperatures is a function of the matrix. The maximum operating temperature of the latter depends on external conditions including environment and stress. Some continuous service temperatures for matrix systems are listed in Chapter 3, Table 3.2. It must be emphasized that for some materials, e.g. epoxy resins, there are many different varieties and that the upper working temperature refers to the system with the highest rating.

Some examples of the mechanical properties of 60 v/o carbon fibre composites at temperature and after exposure to temperature are given in Table 5.21.

Materials

Epoxy, bismaleimide, polyimide and PEEK resins were used, with 60% carbon fibre reinforcement.

Observations

It is clear that polymer matrix composites can retain useful longitudinal properties at high, instantaneous, temperatures and after prolonged periods at elevated temperatures.

Design implications

With care in the choice of materials it is possible to use composites with organic matrices at elevated temperatures.

Table 5.21 Effects of temperature and exposure

property	epoxy	bismaleimide	polyimide	PEEK
instantaneous test at °C		250	343	177
E_{lt} (%)		79	81	
σ_{lt} (%)		71	57	58
after exposure at °C	150	220	260	
time (hours)	30000	10000	8600	
E_{lt} (%)	100	97	71	
σ_{lt} (%)	107	21	63	
test temperature (°C)	150	200	260	

Source

Hancox, N.L. and Phillips, D.C. (1985) *Proc. 2nd Materials Engineering Conference*, London, 139–44.

Resin type and temperature

Further information on the effects of temperature on the properties of unidirectional composite is given in Figures 5.9, 5.10 and 5.11.

Materials

The results are for Toray T300 carbon fibre reinforcement in two or three standard bismaleimide resins (65FWR was referred to in Table 5.18), polystyryl pyridine, a polyimide, PMR15, and an intermediate temperature cure epoxy, Ciba-Geigy's 913C.

Observations

All the results have been expressed as a percentage of the room temperature value. The modulus and strength of the epoxy-based system hold up well up to 150°C, though the interlaminar shear strength begins to decrease immediately the temperature at which the measurements are made increases. The polystyryl pyridine matrix gives the best retention of strength with temperature while the results for the bismaleimides and polyimide-based composites are similar but rather lower.

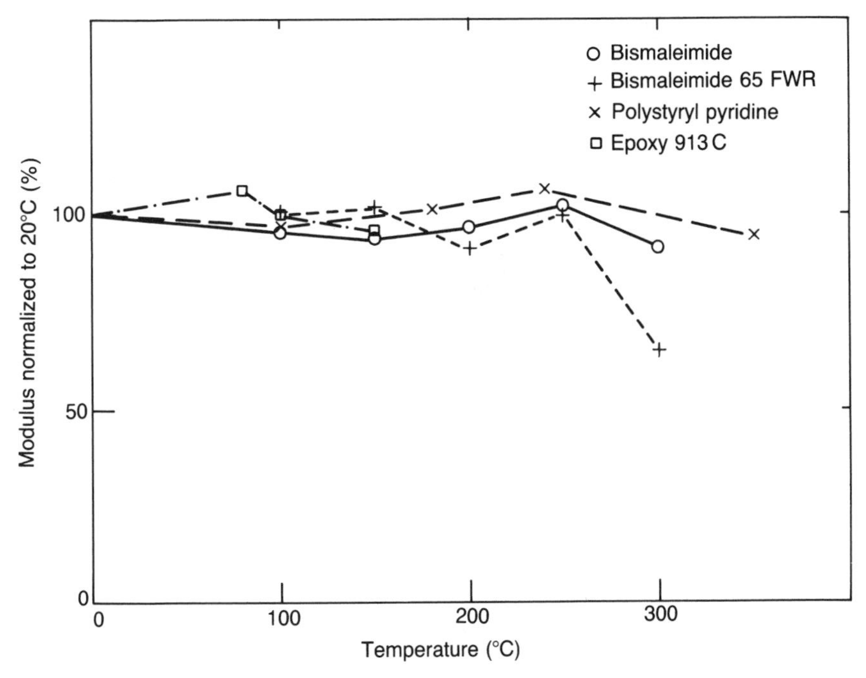

Figure 5.9 Relative modulus versus temperature for carbon fibre composites with different resins. (Reproduced with permission of the Plastics and Rubber Institute, London.)

5.4 CARBON FIBRES

Temperature effects

Resin type

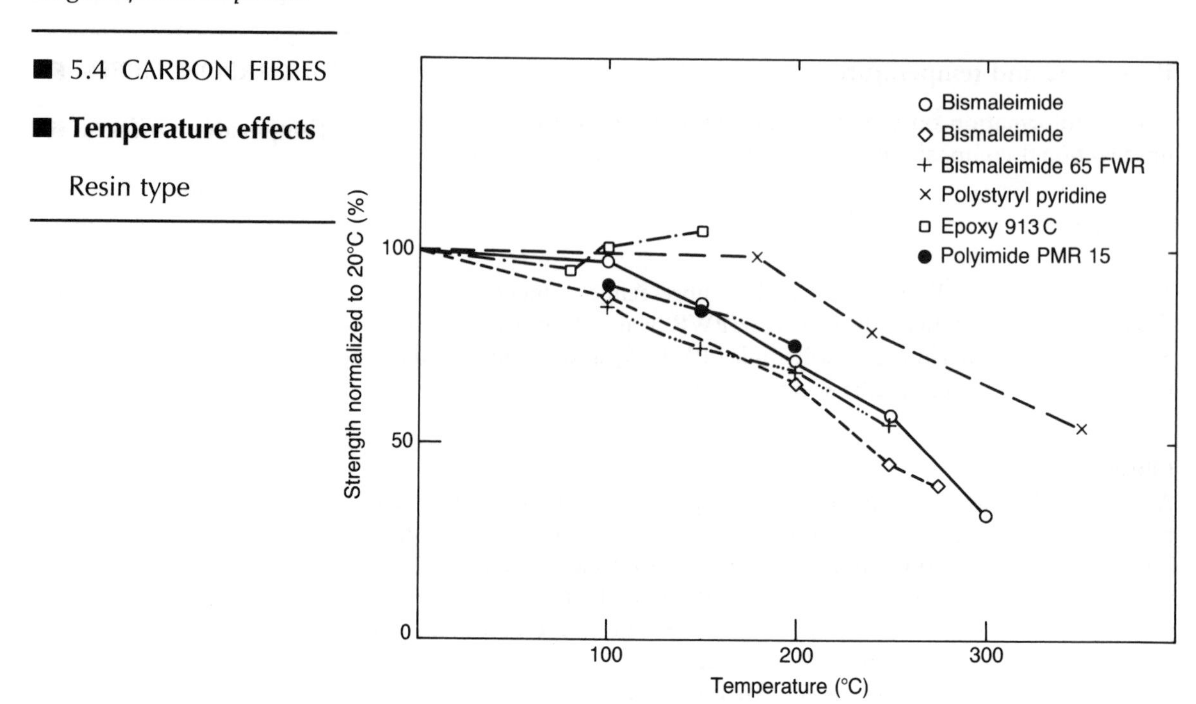

Figure 5.10 Relative strength versus temperature for carbon fibre composites with different resins. (Reproduced with permission of the Plastics and Rubber Institute, London.)

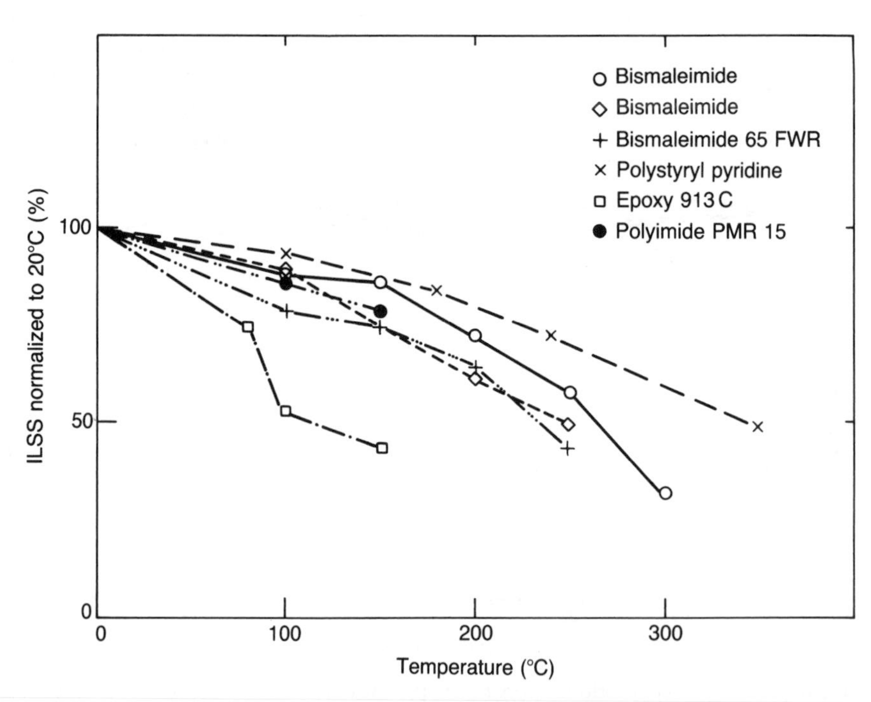

Figure 5.11 Relative ILSS versus temperature for carbon fibre composites with different resins. (Reproduced with permission of the Plastics and Rubber Institute, London.)

Design implications

Stiffness appears to depend less critically on temperature than do strength properties. It must also be remembered that the bismaleimide, polyimide and polystyryl pyridine resins are more expensive than the epoxy and possibly more difficult to process.

Sources

Hancox, N.L. and McLaren, J.R. (1988) *Proc. Conference on Fibre Reinforced Composites*, Liverpool University, 1988.
Plastics and Rubber Institute, London.

Carbon fibre thermoplastic matrix systems

So far when considering carbon fibre composites the matrix has been a thermosetting resin or polymer. The relative merits of thermosetting and thermoplastic polymer matrices were discussed briefly in Chapter 3. Several thermoplastic matrices are compounded with short carbon or glass fibres. In Table 5.22 however the properties of unidirectional, continuous, carbon fibre-reinforced materials are illustrated.

Table 5.22 Thermoplastic matrix carbon fibre composites

property	resin fibre	PAS T650-42	PPS AS4	PEEK AS4	PEEK IM6	APC2
V_f	v/o		66	61	61	
ρ	Mg/m^3		1.6	1.66	1.66	
E_{lt}	GPa	163	135	134	176	139
σ_{lt}	MPa	2300	1656	2130	2700	2131
E_{tt}	GPa			8.9	9.3	10.4
σ_{tt}	MPa			80		64
E_{lf}	GPa		124.2	121	151	
σ_{lf}	MPa		1310	1880	2170	
E_{lc}	GPa					102
σ_{lc}	MPa	1130	655	1100	1100	1116
E_{tc}	GPa					5.7
σ_{tc}	MPa					253
G_{lt}	GPa					4.5
ILSS	MPa		69	105		110

Materials

PAS is an amorphous resin produced by Phillips Petroleum. PPS and PEEK are polyphenylene sulphide produced by Phillips Petroleum and a semi-crystalline polyether ether ketone from ICI, respectively. PEEK is combined at source with a carbon fibre reinforcement to give APC2. Reinforcements are T650-42 carbon fibre, manufactured by Amoco, and AS4 and IM6 carbon fibres made by Hercules.

Observations

It is interesting to compare the efficiency of the reinforcement for AS4 fibre in PPS and PEEK. In the former case the values are 92% and 63% for modulus and strength respectively and in the latter 99% and 87% respectively. Modulus uptake is excellent in both cases and strength in the second. The poorer strength result for PPS may be due to reduced compatibility between the fibre surface and matrix. This is supported by the lower ILSS value for PPS carbon fibre composite. Compressive properties are notably lower than tensile ones, while transverse tensile properties are similar to those of thermosetting polymer matrix materials.

Design implications

As far as short-term mechanical properties are concerned thermoplastic matrix composites have similar tensile properties to those of thermoset systems, though the compressive properties may not be so good. Other factors that have to be taken into account are the relative costs of the systems, ease of fabrication and the impact performance. While it is difficult to generalize, thermoplastic systems may be more expensive and difficult to fabricate than their thermosetting counterparts but may behave better in an impact situation.

Sources

Duthie, A.C. (1988) *Proc. 33rd International SAMPE Symposium*, Society for the Advancement of Material and Process Engineering, 296–307.

Amoco, Phillips Petroleum and ICI trade data.

5.5 MATERIALS SELECTION AND DESIGN STRATEGY

A possible approach to this problem is as follows. The customer/designer must decide on the performance required from the arterfact/structure in terms of short/long-term mechanical, thermal and electrical properties, mass and environmental resistance. From this information he/she can select possible fibre and matrix types, using the information in this book, decide on the lay-up or directional nature of the reinforcement and select a suitable fabrication route. Logically, for a given set of properties, the materials which fulfil the conditions and are the least expensive and easiest to process will be chosen.

It is difficult to be more precise without specific examples though the effects of some limiting conditions are evident: if the structure must be electrically insulating carbon fibres are not suitable (see Chapter 7 for the electrical properties of carbon fibres); if fabrication must be carried out at room temperature a thermoplastic matrix could not be used; while if maximum stiffness or minimum weight is vital, carbon or aramid fibres, respectively, would be employed.

The subject of cost has been deliberately avoided so far since it depends not only on the general type of reinforcement and matrix but on the specific grade and quantity purchased. Furthermore prices are subject to revision, up or down, depending on market forces. Another factor is fabrication, which is so important for composites – it may be better to have a more expensive starting material but a cheaper or more reliable fabrication route.

In *very* general terms the ranking of fibre reinforcement in terms of increasing cost is, glass, aramid, carbon, ceramic. For matrices the ranking, starting with the least expensive, is the group polyester, phenolic, vinyl ester, commodity thermoplastic, followed by epoxies, then bismaleimides and finally polyimides and PEEK. However the only sure way of identifying cost is to consult the appropriate manufacturer.

Further details on composite design are given by Mayer (1993).

5.6 REFERENCE INFORMATION

Jones, R.M. (1975) *Mechanics of Composite Materials*, McGraw-Hill/Kogakusha, Tokyo.

Mayer, R.M. (ed.) (1993) *Design with Reinforced Plastics*, Design Council, London.

6 IMPACT AND FRACTURE

SUMMARY

This chapter considers the impact and fracture behaviour of unidirectional, woven and mat composites. Impact behaviour is a difficult area for the designer and the results given here can only act as an approximate guide to material behaviour and the effects of specimen geometry and edge conditions.

6.1 INTRODUCTION

The impact and fracture behaviour of composites is complex. Thermosets do not undergo plastic deformation, composites are not necessarily described by linear elastic fracture mechanics (LEFM) and it is very difficult, if not impossible at the present time, to predict the initiation and progression of failure in complicated structures from the behaviour of unidirectional materials. Because of the latter much of the work on fracture has involved studies of plied or laminated specimens.

There are several approaches to impact and fracture failure and it is useful to allude to some of them before giving specific data.

One way of rating materials is based on the elastic energy that can be stored, in a unit volume, for a particular mode of stressing. For the tensile case this is proportional to

$$\frac{\sigma_{lt}^{2}}{E_{lt}}$$

hence materials with a high tensile strength and low modulus have good energy absorbing properties. Unfortunately this method does not take into account damage initiated internally before final failure which may seriously degrade the properties of the component.

The micromechanics approach sums the energy absorbed by various mechanisms occurring in the total process including stored elastic energy and that absorbed in fibre pull-out, debonding, etc. Fracture energy measurements involve loading specimens in the I, II or III fracture modes (see Figure 6.1) or a mixed mode, and measuring G_{IC}, G_{IIC} or G_{IIIC} for both initiation and propagation of a crack, though unfortunately it can sometimes be extremely difficult to interpret the experimental data. Provided the material obeys LEFM, i.e. the process taking place at the crack tip can be realistically analysed in terms of linear elasticity theory, it is possible to derive a stress intensity factor, K, from the data. The situation is more complex in laminates than in unidirectional specimens since

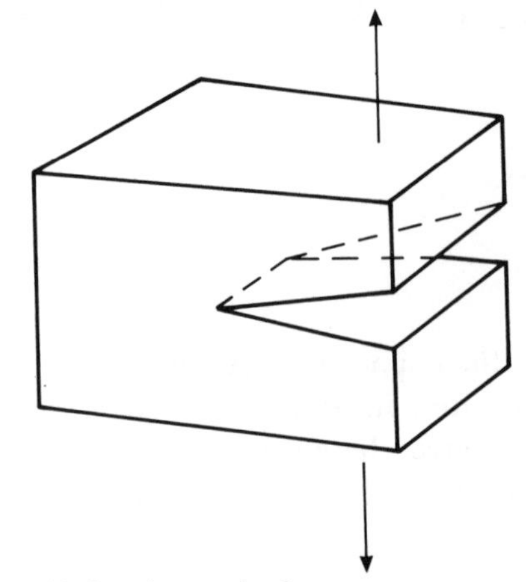

(a) Opening mode, G_I

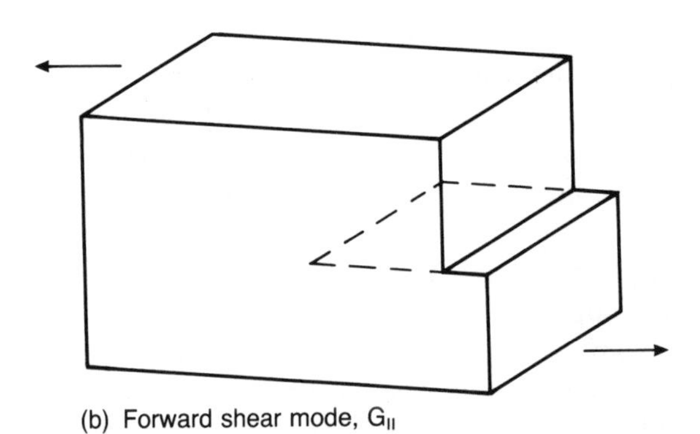

(b) Forward shear mode, G_{II}

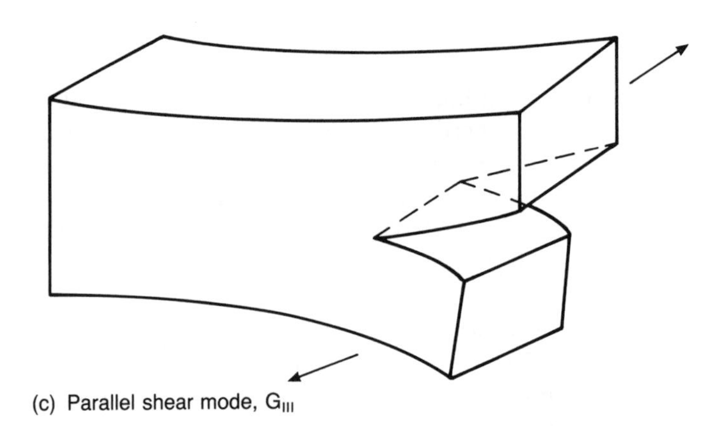

(c) Parallel shear mode, G_{III}

Figure 6.1 Crack extension modes.

delamination can occur between plies in addition to other failure processes. A complete discussion is given by Friedrich (1989). Another cause for concern in real structures is barely visible impact damage (BVID). Any of the above processes may be modified by the application of an external stress – the effect of impact and tension combined can be much more serious than impact alone.

Various methods are used to assess fracture behaviour. Charpy and Izod pendulum tests are of more use for metals or homogeneous materials or for determining the effect of temperature on the ductile/brittle transition, though instrumented Charpy or drop weight tests are used sometimes to assess composites. In instrumented drop weight tests the indenter can be chosen to be similar to that which might be encountered in practice, whilst the impact resistance can be determined from the degree of penetration. Another method measures a property after an impact of specified energy. The variable studied is the one regarded as critical for the particular application. This may be the leakage pressure of a tube or the compression strength of a panel. The details of the latter, DIN 65560, are in Mayer (1993). Fracture energies are measured in a wide variety of test procedures. Alternatively relative or comparative information may be derived for specific lay-ups, experimental geometries and impact conditions. Although not fundamental such information is useful for ranking materials in a particular situation.

In practical applications the designer or analyst may combine a failure criterion and knowledge of known defects and fracture energies to predict how and where failure is initiated and the subsequent progress of damage in the material. Such analyses usually have to be augmented by empirical data to model the complete failure process.

The impact performance of unreinforced polymers is very difficult to quantify. The impact strength is usually determined in an Izod or Charpy test. Different impact tests on the same material (Figure 6.2) do not give the same numerical result and the designer has to choose the test and test data most appropriate to that encountered in service. In addition results will depend on the temperature and possibly on the presence or otherwise of a notch and its radius of curvature, the moisture content of the material, the overall dimensions of the specimen, the molecular weight of the polymer, the presence of additives, etc. At room temperature some thermoplastics, e.g. polypropylene, polyamide and polysulphone, are brittle if suitably notched, while wet polyamide is not so and is said to be tough.

An illustration of the effect of blending an iso-polyester resin with a urethane acrylic is shown in Figure 6.2. The beneficial effect of adding a flexible resin on the impact strength is clear. The effects of the additive on tensile properties are shown in Figure 4.1.

The fracture energy, G_{IC}, of an epoxy resin can range from 50 to 11,500 J/m^2 for a toughened system. However other properties of the highly toughened material such as its upper working temperature, modulus

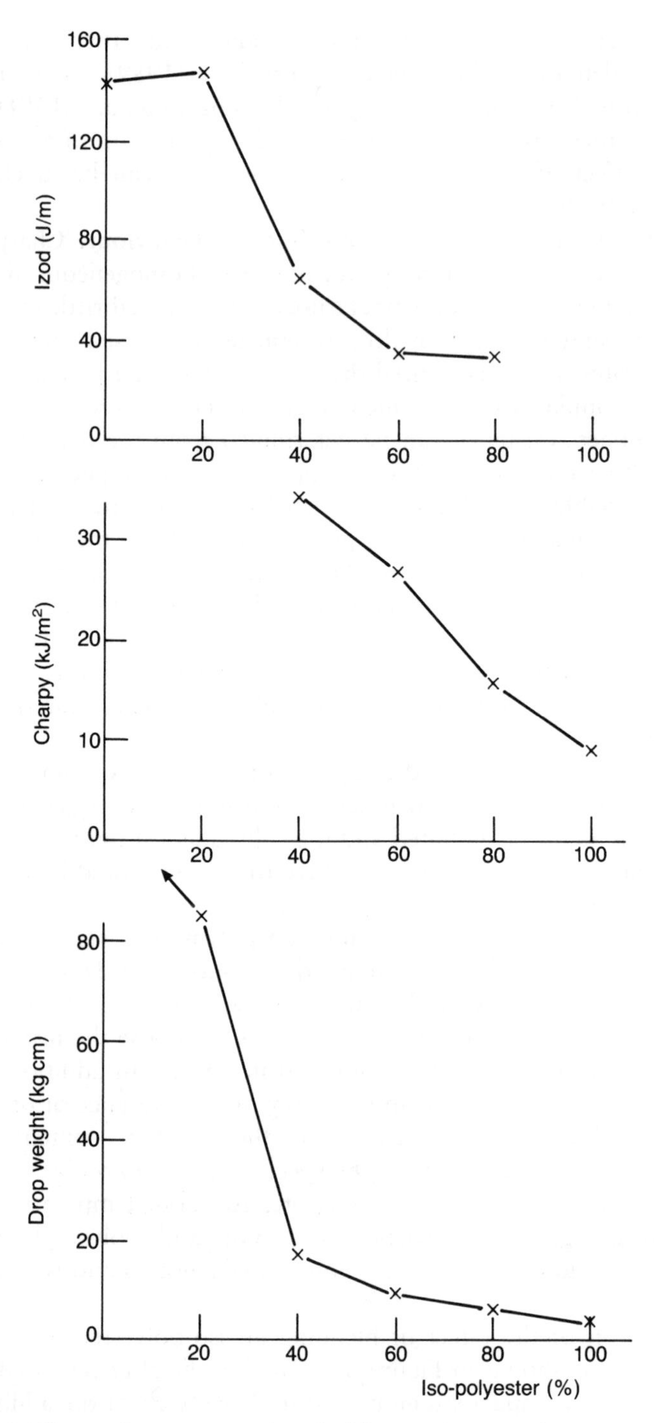

Figure 6.2 Effect on impact properties of adding urethane acrylate to cast iso-polyester resins (Norwood).

and strength, may be so reduced as to make it effectively useless as a matrix for a composite system. The room temperature work of fracture of high performance thermoplastics, which may be used in conjunction with a fibrous reinforcement, ranges from 650–6600 J/m^2.

Since the subject of this book is composites and fibres are usually added to resins to increase stiffness and strength, rather than deliberately to alter impact or fracture properties, it is not proposed to give further details of the impact performance of unreinforced polymers here.

Source

L. Norwood at Scott Bader.

■ 6.2 IMPACT STUDIES

■ **Charpy impact**

6.2 IMPACT

The force displacement characteristic generated by an instrumented falling weight or swinging pendulum striking a specimen rises to a peak value and then decays to zero. The energy stored up to the peak force is often referred to as the initiation fracture energy and the area under the whole curve as the total impact energy or toughness. This view is rather simplistic since localized or microfailure may be initiated prior to the peak load and the total energy involved in the process may include contributions associated with the equipment and specimen geometry rather than the true failure process. One method of reducing the influence of geometry is to notch the specimen so that the impact energy is used to cause fracture and excess energy is not stored throughout the specimen prior to failure. However notch design and the state of stress at the tip of the notch are complex issues.

Some data for the Charpy impact of glass composites, using 3- or 4-point loading and notched and un-notched specimens, are given in Table 6.1. The data for 3-point Charpy loading are also shown in

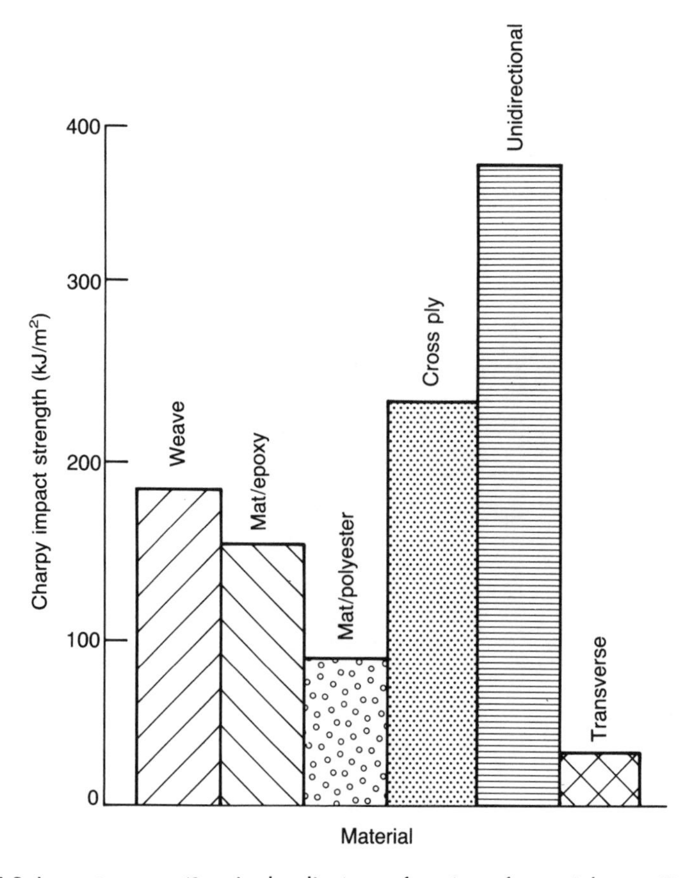

Figure 6.3 Impact energy (3-point loading) as a function of material type. (Reproduced with permission from Sims, G.D., *Proc. ICCM 6*; 3.494–3.507; published by Elsevier Applied Science, 1988.)

6.2 IMPACT STUDIES

Charpy impact

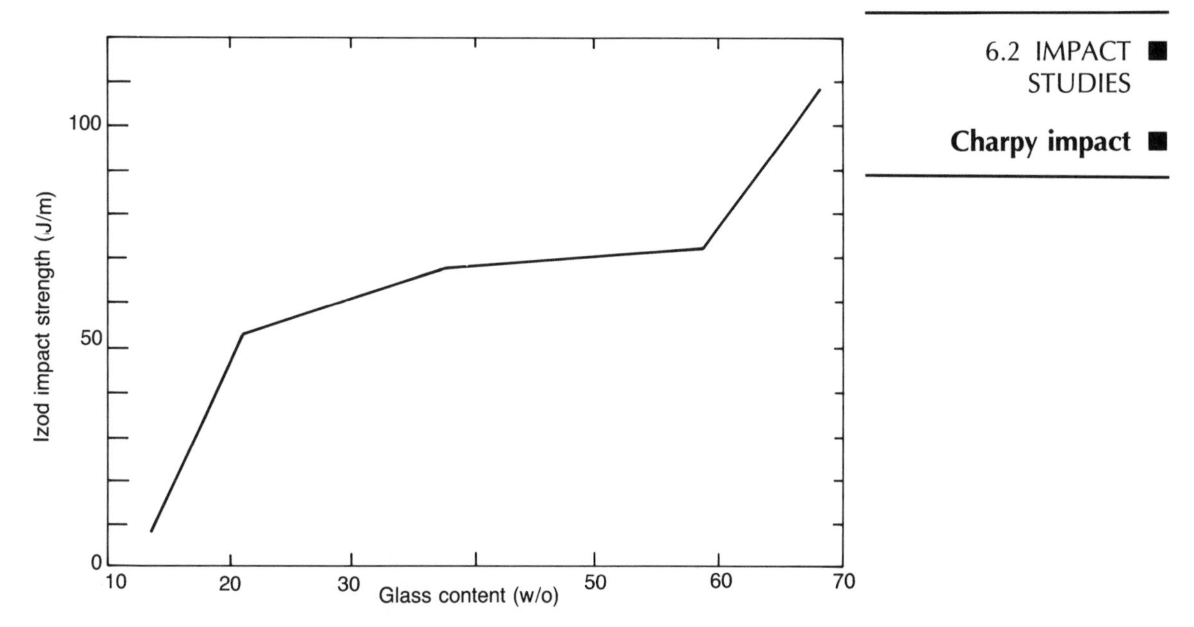

Figure 6.4 Effect of glass content as chopped strand mat on Izod impact strength. (Redrawn with permission from Orpin, M., BP Chemicals Ltd.)

Figure 6.3. Izod impact data for glass fibre chopped strand mat phenolic specimens with varying amounts of glass fibre are shown in Figure 6.4.

Materials

The reinforcement is E-glass in the form of chopped strand mat, woven rovings, fine weave fabric and unidirectional fabric; and the resins were epoxies, phenolics (BP Chemicals), polyesters or compounds (DMC/polyester, SMC/polyester).

Observations

The first six results in Table 6.1 are for an impact velocity of 2.44 m/s. The impact energy or toughness depends on the mode of stressing, the presence or otherwise of a notch, V_f, and the material disposition. Particularly noticeable are the low results for the 20 v/o mat specimen and the transverse unidirectional material, where there is no reinforcement to increase toughness.

For moulding compounds the difference between the values for SMC and DMC is due to the presence of longer fibres in the former case. Overall the SMC values compare favourably with those of mat polyester. The materials with continuous fibre reinforcement have much higher impact properties provided a reasonable proportion of the fibres is in the direction in which the main bending stresses are generated in the impact test.

The Charpy impact strength for 60 v/o epoxy Kevlar 49 unidirectional composite is 240–310 kJ/m^2, similar to the figures recorded for unidirectional GRP. Usually information for aramid composites refers to

6.2 IMPACT STUDIES

Charpy impact

Table 6.1 Charpy impact data for various types of GRP laminate

	V_f (v/o)	impact energy (kJ/m^2)		
		3-point	4-point	notched
woven roving/ polyester	38	175	214	146
CSM mat/epoxy	42	143	225	114
CSM mat/polyester	20	81	94	54
balanced cross ply/ epoxy	61	227	214	229
UD/epoxy longitudinal	61	360	609	413
UD/epoxy transverse	61	28	26	8.8
DMC/polyester		13–25	13–25	
SMC/polyester		55–90	55–90	19–38
SMC/polyester		65	65	
CSM mat/phenolic	20	75	75 (Izod)	

aramid glass or aramid carbon hybrids, since the material is often used in this manner to overcome the low compression strength of aramid fibre composites.

Design implications

Toughness varies markedly. In many cases it is better to test the final design in impact rather than try to predict behaviour on the basis of coupon tests.

Sources

Sims, G.D. (1988) *Proc. International Conference on Composite Materials* 6, 3.494–3.507, Elsevier Applied Science, London.
M. Orpin at BP Chemicals.
DSM Resins and BIP Chemicals Ltd.

6.2 IMPACT STUDIES

Charpy impact

Resin type

Charpy impact studies

Table 6.2 shows results for aligned and cross ply laminates. The notched specimens were tested in 3-point loading in a Charpy machine at an impact velocity of 3.16 m/s.

Materials

The two epoxy resins, which are a temperature and an impact resistant system respectively, are made by Ciba-Geigy, the polyimide, Kerimid 601, by Rhône Poulene, the PES and PEEK by ICI and the PEI by General Electric. Designation (1) and (2) refer to solution and melt impregnation respectively. The reinforcement was XA-S carbon fibre (Courtaulds). All the specimens, except the 914C epoxy ones, had a fibre loading of 50 v/o of Courtaulds XA-S carbon fibre. The 914C specimens had a 60 v/o fibre loading.

Observations

The results in Table 6.2 for CFRP are lower than those in Table 6.1 for GRP because of the lower bending strain energy in the former case. Comparing the impact performance of longitudinal and transverse fibre orientations shows that both CFRP and GRP are highly anisotropic.

Table 6.2 Charpy impact data for carbon fibre composites

material	direction/lay-up	impact energy (kJ/m^2)
epoxy 914C – temperature resistant (Ciba-Geigy)	longitudinal	20
	transverse	0.25
	cross ply	5
epoxy 920C – impact resistant	longitudinal	20
	transverse	1
	cross ply	4
polyimide – Kerimid 601 (Rhône Poulenc)	longitudinal	71
	transverse	0.25
	cross ply	50
PES(1) – solution impregnated (ICI)	longitudinal	76
	transverse	2.5
	cross ply	50
PES(2) – melt impregnated (ICI)	longitudinal	110
	transverse	2.5
	cross ply	75
PEI (General Electric)	longitudinal	26
	transverse	2
	cross ply	38
PEEK (ICI)	longitudinal	73
	transverse	2
	cross ply	40

■ 6.2 IMPACT STUDIES

■ **Charpy impact**

Resin type

The transverse impact strength is low as there are few, if any, fibres in this direction to contribute to strength and hence impact resistance. The importance of fabrication is shown by the results for PES(1) and PES(2) – melt impregnation presumably gives a better composite. The impact results for PEI are very low for a thermoplastic and again probably indicate the effects of fabrication.

Source

Stori, A. and Magnus, E. (1983) Paper 24, in *Composite Structures*, (ed. Marshall, I.H.), Elsevier Applied Science, London.

The effect of reinforcement type on impact behaviour

The results for dropweight impact given in Figure 6.5 show the effect of varying the fibre type on the peak impact force and overall deflection, when other variables including strike velocity, impactor diameter, sample thickness and support diameter are kept constant.

Materials

The fibres are E- and R-glass and Courtaulds EXA-5 carbon fibre. In all cases the lay-up is 8-plies $[0_{90} \pm 45]_S$, and the approximate fibre loading 50 v/o. For the hybrid construction the E-glass layers were on the outside. The resin matrix was ICI Modar, a modified methacrylate.

Observations

This is a set of consistent data from instrumented drop weight experiments. The strike velocity was 6 m/s, the impactor diameter 20 mm, the specimen thickness 3.2 mm and the support diameter 50 mm. The

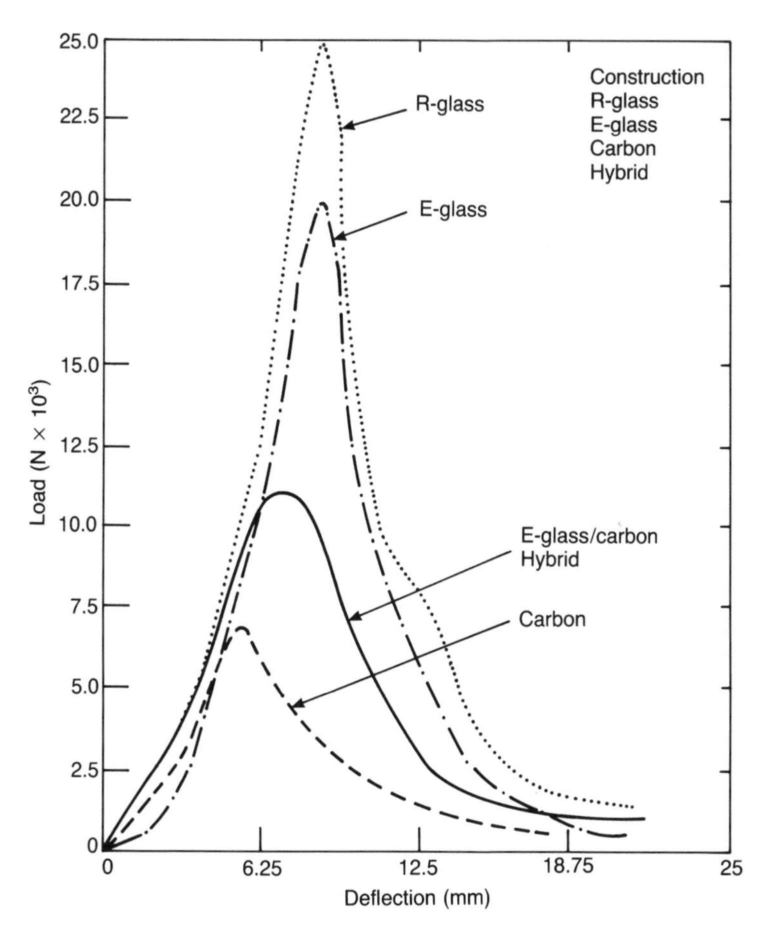

Figure 6.5 The effect of fibre type and hybrid construction on impact. (Reproduced with permission of Walker, R.A., PERA International (UK).)

■ 6.2 IMPACT STUDIES

■ **Reinforcement type**

initiation energy, defined as the area under the load/deflection curve up to peak load, is highest for the R-glass composite and lowest for the carbon fibre material. The impact performance of R-glass composite is superior to that of E-glass composite and both are superior to CFRP. This occurs because of the relative values of the stored strain energy per unit volume; for tensile loading the strain energy is proportional to $\left(\frac{\sigma_{lt}^2}{E_{lt}}\right)$. A similar relationship holds for flexural loading, neglecting the influence of shear. Hybrids, i.e. a combination of glass and carbon fibres in a common matrix, have an impact strength between that of glass and carbon composites.

Design implications

These data support the view that a combination of high strength and low modulus, and hence high stored strain energy, gives a good impact performance. In addition the combination of different fibres permits high stiffness with good impact properties. This approach is often used in the aerospace industry.

Source

Impact database, PERA International, UK.

Strain rate

Chopped strand mat polyester

The effects of strain rate on impact

The effects of increasing the strain rate over five decades is shown in Figure 6.6. Parameters such as specimen thickness are constant.

Material

The specimens were made from E-glass chopped strand mat and polyester resin with a fibre volume loading of 20 v/o.

Observations

The maximum load and the area under the load deflection curve increase with strain rate. Consequently the energy to initiate fracture and the total fracture energy also increase with strain rate.

Design implications

GRP materials become stronger and tougher as the strain rate increases.

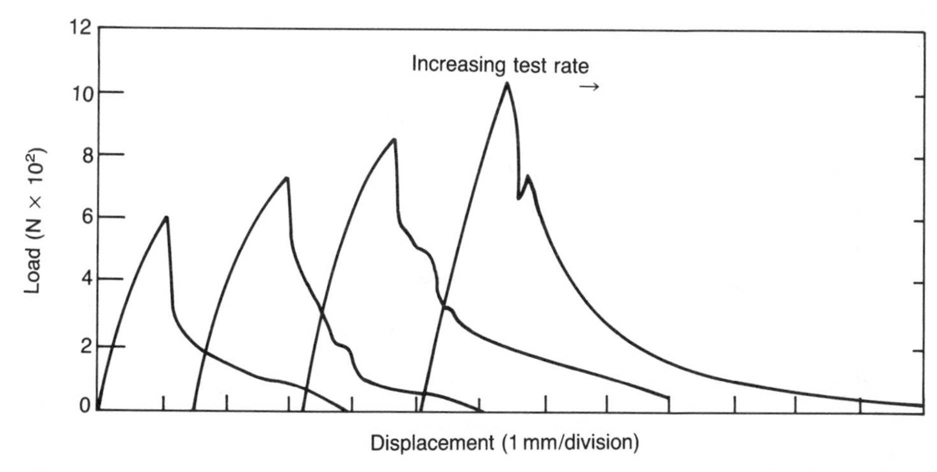

Figure 6.6 Load/displacement curves for mat polyester specimens at 10^{-6}, 10^{-4}, 10^{-2} and 10^{-1} m/s displacement rate. (Reproduced with permission from Sims, G.D., *Proc. ICCM 6*; 3.949–3.507; published by Elsevier Applied Science, 1988.)

Source

Sims, G.D. (1988) *Proc. International Conference on Composite Materials* 6, 3.494–3.507, Elsevier Applied Science, London.

■ 6.2 IMPACT STUDIES

■ **Loading rate**

The effects of loading rate on impact

Other relevant results are given in Figure 6.7, which shows the ultimate tensile strength as a function of loading rate, and in Figure 10.20, where the effect of loading rate on fatigue strength is illustrated. It is clear that increasing the loading rate increases the tensile and fatigue strengths significantly.

Materials

The resin was an epoxy, Permaglass 22 FE from Fothergill and Harvey, the reinforcement was E-glass fibre woven fabric.

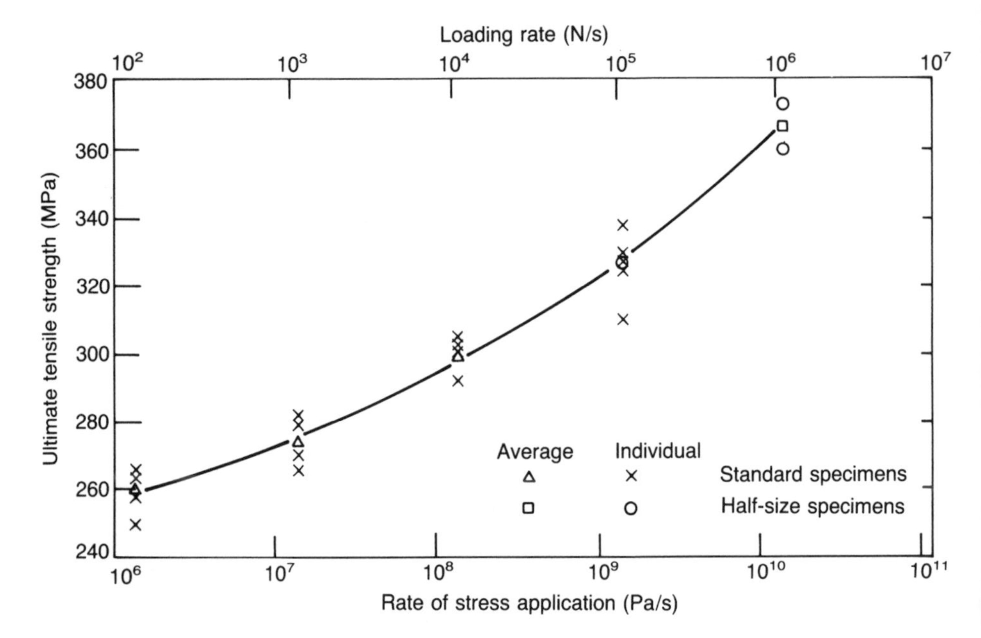

Figure 6.7 Ultimate tensile strength as a function of loading rate. (Reproduced with permission of Sims, G.D. and Gladman, D.G.)

Source

Sims, G.D. and Gladman, D.G. (1982) *DMA (A) 59*, National Physical Laboratory, Teddington.

6.2 IMPACT STUDIES

Support diameter

The influence of support diameter on impact properties

Figure 6.8 shows the effect of varying the support diameter on specimens struck in a drop weight impact test.

Material

The specimens were made from 450 g/m^2 E-glass chopped strand mat with a Scott-Bader isopolyester resin, Crystic 272. The fibre volume fraction was 36 v/o giving a specimen thickness of 5 mm.

Observations

Although the peak load is constant the work of fracture and deflection increase with support diameter. Other results show that provided the support diameter is greater than 100 mm the peak force in the test is least for the 36 v/o CSM and greatest for the 55 v/o woven E-glass roving specimens with figures for the 46 v/o woven roving/CSM material in between. In all cases the resin is Crystic 272.

Design implications

Not only do the material and impact velocity affect impact behaviour, but so does the way the specimen is supported since this influences the way it bends when struck.

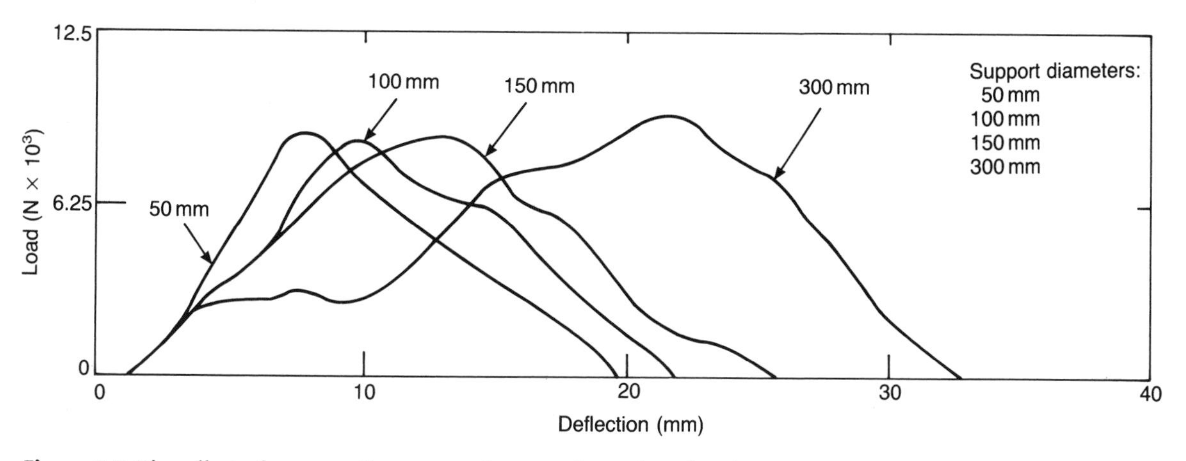

Figure 6.8 The effect of support diameter on impact. (Reproduced with permission of Walker, R.A., PERA International (UK).)

Source

Impact database, PERA International, UK.

■ 6.3 FRACTURE BEHAVIOUR

■ **Fracture energy**

6.3 FRACTURE

A better way, in principle, of specifying fracture behaviour is to measure the energy, G_C, required to create two new surfaces in the material. If LEFM applies this information can be related to a stress intensity factor (K). G_C may depend on whether it refers to initiation or propagation, crack length and velocity, the stressing mode and direction in the material. In addition it is necessary to avoid other routes for energy loss in the experimental determination. Because of its definition values of G_C are usually much smaller than the impact energies determined in drop weight or pendulum tests.

Some values of G_{IC} and K_C are given in Table 6.3. See Figure 6.1 for a description of the loading involved for G_{IC}.

Materials

Kevlar 49 is an aramid fibre made by du Pont, T300 carbon fibre is made by Toray and AS1 and AS4 carbon fibres by Hercules. XA-S is a Courtaulds carbon fibre, and E-glass cloth, 50 v/o, was also used. The epoxies are a variety of aerospace grades: F155 (Hexcel); PR286 (3M); 5208 (Narmco); 3501-6, 3502 (Hercules); 914 (Ciba Geigy); 934 (Fiberite). APC1 and APC2 are based on carbon fibres in a PEEK matrix (ICI).

Table 6.3 Fracture energies and stress intensity factors

material	G_{IC} (J/m^2)	K_C ($MN/m^{3/2}$)
Kevlar 49, 181 fabric, 50 v/o, $[\pm 45]_s$, Hexcel F-155 epoxy		17.4–27.2
as above, $[0, 90, \pm 45]_s$		21.8–32.4
Kevlar 49, unidirectional tape, 60 v/o, $[\pm 45]_s$, 3M PR286 epoxy		11.1–18.2
as above, $[0, 90, \pm 45]_s$		26.4–40
unidirectional E-glass epoxy	525–1020	
E-glass cloth fabric, 50 v/o epoxy	1550–1990	4.6–5.4
carbon fibres (T300, AS1, AS4), epoxies (5208, 3501–6, 3502, 914, 934)	60–254 average 145	
T300 carbon fibre, 934 epoxy	530–690	5.50–6.84
carbon fibre, epoxy	203 CV 18% 115 (initiation) CV 15%	
carbon fibre, PEEK	1707 CV 8% 1406 (initiation) CV 15%	
XAS carbon fibre, PEEK (APC1)	1408	
AS4 carbon fibre, PEEK (APC2)	1330–2890 average 1990	

Note: CV refers to cofficient of variation.

Observations

The data have been determined by a variety of test methods, for specimens with different geometries. The carbon fibre epoxy and PEEK data are based on a considerable number of experiments in a 'round robin' programme. The range of K_C data for Kevlar 49 materials is due to different crack lengths as is the range in the property for T300 carbon fibre composite. There is some evidence that the initiation fracture energy in unidirectional carbon fibre epoxy composites is considerably lower than the value quoted here. The G_{IC} values for thermoplastics are considerably higher than for epoxy matrix composites. The effect of temperature is not very marked. Tung and Liao investigated this for unidirectional AS4/3501-5A, carbon fibre epoxy noting that G_{IC} varied from 200 to 219 J/m^2 over the range (−100 to 100°C).

Design implications

Great care must be exercised when considering the fracture aspects of a design.

Sources

Bathias, C. and Laksimi, A. (1985) *ASTM STP 876*, ASTM, 217–37.
Davies, P. *et al.* (1992) *Composites Science and Technology*, **43**, 129–36.
Saghizadeh, H. and Dharan, C.K.H. (1986) *Journal of Engineering Materials and Technology*, **108**, 290–5.
Sela, N. and Ishai, O. (1989) *Composites*, **20**, 423–35.
Tung, C.M. and Liao, T.T. (1989) *SAMPE Quarterly*, **20**(3).
Ye, L. (1992) *Composites Science and Technology*, **43**, 49–54.
du Pont trade literature.

■ 6.3 FRACTURE BEHAVIOUR

■ **Fracture energy**

Fracture energy

Table 6.4 gives some G_{IIC} data.

See Figure 6.1 for description of G_{IIC}. For APC2 refer to Table 6.3.

Materials

5208, 914, 934 and 3501-6 epoxy resins were used (Narmco, Ciba Geigy, Fiberite and Hercules, respectively) as was PEEK (ICI). Reinforcements were E-glass fibre, T300 carbon fibre (Toray) and AS4 carbon fibre (Hercules).

Observations

G_{IC} and G_{IIC} increase as one goes from carbon fibre epoxy to glass fibre epoxy to carbon fibre PEEK.

Table 6.4 G_{IIC} data for composites

material	G_{IIC} (J/m^2)
T300, AS4 carbon fibre and epoxies (5208, 914, 934, 3501–6)	154–518 average 450
carbon fibre, epoxy	437 CV 14%
APC2	1930
AS4 carbon fibre, PEEK	1109–1860
carbon fibre, PEEK	2033 CV 23%
E-glass fibre, epoxy	1200

Sources

Davies, P. *et al.* (1992) *Composites Science and Technology*, **43**, 129–36.
Sela, N. and Ishai, O. (1989) *Composites*, **20**, 423–35.
Takeda, N., Tohdoh, M. and Takahashi, K. (1989) *SAMPE Quarterly*, **20**(2), 27–32.

6.3 FRACTURE BEHAVIOUR

Fracture energy

Strain rate

The effects of strain rate on fracture energy

The effects of strain rate are shown in Table 6.5.

Materials

Thornel is made by Amoco, AS4 by Hercules and the aramid fibre, Kevlar 49, by du Pont. 934 and 3501-6 are aerospace epoxy resins from Fiberite and Hercules, respectively.

Observation

A thousand-fold increase in strain rate increases G_{IC}, for one of the systems, by 28%. The trends of the results are similar to those in Figure 6.6 for glass mat polyester composite.

Design implications

Most real life impact situations correspond to relatively high strain rates whereas many measurements are made at low strain rates. Hence measured data may be conservative.

Table 6.5 The influences of strain rate on G_{IC}

material	strain rate (mm/s)	G_{IC} (J/m^2)
carbon fibre, Thornel 300/934, unidirectional	0.083	90
as above, woven	0.083	232
	0.83	266
woven Kevlar 49/934	0.083	330
	0.83	382
carbon fibre AS4/3501–6, unidirectional	0.0085	198
	0.85	222
	8.47	254

Sources

Aliyu, A.A. and Daniel, I.M. (1985) *ASTM STP 876*, ASTM, 336–48.
Saghizadeh, H. and Dharan, C.K.H. (1986) *Journal of Engineering Materials and Technology*, **108**, 290–5.

6.4 DESIGN STRATEGY

The impact fracture performance of composites has always been a critical and contentious issue. Standard test methods are still not agreed, though progress is being made, and the interpretation of the basic test data is open to argument particularly with reference to initiation and propagation energies. Although it may be possible to account for a specific value of G_C or impact energy in terms of micromechanical energy absorption methods, it is generally not possible to take fracture data for a particular material and calculate the fracture behaviour of a structure. The information given here helps to rank different materials and constructions and clarify the effects of geometrical variables. A suggested strategy for designing an impact resistant structure is based on the following squence; fibre length, fibre volume loading, fibre lay-up, fibre type and resin type. However, the best approach to assessing failure behaviour is still to fabricate a full size component and test it.

6.5 REFERENCE INFORMATION

Friedrich, K. (ed.) (1989) *Application of Fracture Mechanics to Composite Materials*, Volume 6 of Composite Materials Series, Elsevier, Amsterdam.

Mayer, R.M. (ed.) (1993) *Design with Reinforced Plastics*, Design Council, London.

Impact database – a database has been developed by PERA to facilitate the computer-aided design of impact-resistant composite structures. For further information contact PERA International, Melton Mowbray, Leicestershire LE13 0PB, UK. Telephone (44) 0664 501 501.

7 THERMAL AND ELECTRICAL PROPERTIES

SUMMARY

The thermal properties of fibre-reinforced composites are anisotropic. Expansion in the direction of the fibre is usually very small or negative, but the thermal conductivity of some carbon fibre composites in the fibre direction can be high. Most polymers and several types of fibre are good electrical insulators. Carbon and some ceramic fibres have a much lower resistivity. It is thus possible to use composites for manufacturing radomes as well as for electromagnetic screening materials and systems which absorb radar signals.

7.1 THERMAL PROPERTIES

The coefficients of thermal expansion (CTE) and thermal conductivity (CTC) of unidirectional composites may be greatly influenced by the reinforcement and are highly anisotropic. Most information pertaining to the thermal properties of fibres, other than the temperature of oxidation or degradation, is obtained from measurements on composites using a model relating composite properties to those of the individual phases, their volume fractions and their properties. Generally the bulk organic matrices are isotropic and have a high CTE ($30-100 \times 10^{-6}$/°C) and low CTC (<0.5 W/m.K). The CTE is essentially constant up to the second order glass transition temperature, T_g, and then increases (up to 200×10^{-6}/°C).

An analysis of measurements on composites indicates that some carbon and aramid fibres have a small negative CTE (-0.1 to -3.5×10^{-6}/°C) along the long fibre axis and a much larger, positive, CTE (up to 40×10^{-6}/°C) in the radial or perpendicular direction. Glass and aramid fibres have a very low CTC but that of carbon fibres in the direction of the longitudinal fibre axis can, depending on the type of fibre, be one or two orders of magnitude greater (up to several hundred W/m.K) and may increase with temperature up to possibly 100°C. The highest CTC is found for the highly internally aligned, very high modulus, carbon fibres. A few direct measurements of fibre CTC have been reported at temperatures approaching absolute zero. Transverse to the long fibre axis the CTC of a composite is much lower (up to 1 W/m.K) and very dependent upon the matrix, though fibre misalignment or bridging in an otherwise unidirectional composite may increase the conductivity slightly.

The thermal performance of a polymer is often specified in terms of its glass transition temperature, T_g. This is the temperature at which the polymer undergoes a transition from a glass to a rubber. Several properties change significantly at, or in the vicinity of, T_g – e.g. the CTE of the polymer increases markedly on going from the glass to the rubber state. For further details see Young (1981). Another, empirical, means of defining the thermal performance of a polymer is the heat distortion temperature, HDT. This is the temperature at which a simple rectangular cross-section beam of the polymer, with a load applied at the centre to give a maximum bending stress of 0.455 or 1.82 MPa, deflects by 0.25 mm when heated in an oil bath at 2°C/min. Further details are given in ASTM D 648. The continuous service temperature, referred to in Chapter 3, is more difficult to assess since it depends on applied stress and environment. All three measurements are influenced by the chemical nature, molecular structure and, if appropriate, degree of cure of the polymer and presence of a filler or reinforcement.

Thermal properties of polymer matrix systems

Some general thermal properties of typical polymer matrices are given in Table 7.1.

Materials

The resins used were: isophthalic polyester; phenolic; epoxy; bismaleimide; and several thermoplastics – PES, PEEK (ICI) and PPS (Phillips Petroleum).

Observations

The figures quoted are for measurements at room temperature or slightly above but below the T_g of the system. The specific heat and particularly the CTE increase with increasing temperature.

Table 7.1 Thermal properties of polymers

type	CTE ($\times 10^{-6}\,°C^{-1}$)	CTC (W/m.K)	C_p (J/kg.K)
polyester (isophthalic)	100–110	0.08–0.2	960–1250
phenolic	10–20	0.2–0.25	2000
epoxy	44–120	0.21	1400
bismaleimide	31–49		
PES	55	0.26	1000
PPS	49	0.288	1090
PEEK	40–47	0.25	1340

Sources

Trade data.

■ 7.1 THERMAL PROPERTIES

■ **Fibres**

Thermal properties of fibres

Some quoted data on the thermal expansion and thermal conductivity of fibres are given below in Table 7.2.

Materials

Full details of the reinforcement fibres are given in Table 7.2.

Observations

The anisotropy of the basic carbon and aramid fibres is clear as is the high longitudinal thermal conductivity of the very high modulus Amoco carbon fibres.

Design implications

It is possible to use a simple rule of mixtures to estimate longitudinal thermal conductivity given that of the fibre. Although many models are

Table 7.2 Thermal properties of fibres

type	CTE_l ($\times10^{-6}$ °C^{-1})	CTE_t ($\times10^{-6}$ °C^{-1})	CTC_l (W/m.K)	CTC_t (W/m.K)
carbon fibre				
T300 (Toray)			8.9	
HTA (Akzo)	−0.1		17	
STA (Akzo)	−0.1		17	
HM35 (Akzo)	−0.5		115	
XA (Courtaulds)	−0.26	26	24	
HM (Courtaulds)	−1.3	25	105	
P75 (Amoco)	−1.46	12.5	190	2.4
P100 (Amoco)	−1.48	12.0	536	2.4
P120 (Amoco)	−1.5	12.0	610	2.4
aramid fibre				
Kevlar 49 (du Pont)	−2.33	41.4		
Twaron (Akzo)	−3.5		0.05	0.04
glass fibre				
E (Vetrotex)	5		1	
D (Vetrotex)	2–3		0.8	
R (Vetrotex)	4		1	
S-2 (Owens Corning Fiberglas)	2.3			
ceramic				
polycrystalline metal oxide				
Nextel 312 (3M)	3			
Nextel 440 (3M)	4.38			
Nicalon SiC (Nippon Carbon)	3.1		11.6	
Tyranno SiC (Ube Industries)	3.1			

available the accurate prediction of the transverse coefficient of thermal conductivity is much more difficult. Equations are available to calculate the expansion properties of a unidirectional composite from the thermal and elastic properties of the fibre.

Sources
Trade data.

■ 7.1 THERMAL PROPERTIES

■ **Expansion**

Thermal expansion properties of composites

Materials

The thermal expansion properties of various composites and the manufacturers of the different fibres and matrices are given in Table 7.3. Note that Courtaulds have ceased making carbon fibres but information on their material has been included here since other manufacturers produce fibres with similar properties, see Figure 3.1, and the thermal properties of Courtaulds' materials may be relevant to these other fibres.

Observations

The extreme thermal expansion anisotropy and very low longitudinal expansion of the carbon and aramid fibre systems are clear. All measurements are for room temperature or slightly above. Once the T_g of the polymer matrix has been exceeded the CTE in either direction will increase.

The longitudinal CTE is dominated by the fibre properties and is not very sensitive to fibre volume fraction. The transverse CTE is also reasonably insensitive to fibre volume fraction because both components have high individual expansions in this direction.

Design implications

Care must be taken when using unidirectional laminates because of the large differences in thermal expansion parallel and perpendicular to the fibres. Using laminate plate theory it is possible to design composites based on carbon or aramid fibres with some interesting in-plane expansion properties. However through-thickness expansion will still be large, similar to the transverse values in Table 7.3. The anisotropy of the coefficient of thermal expansion is important when designing joints, particularly when joining dissimilar materials (e.g. composites and metals).

Sources

Barnes, J.A., Simms, I.J., Farrow, G.J. *et al.* (1990) *Journal of Thermoplastic Composite Materials*, 3, 66–80.

Bowles, D.E. and Tomkins, S.J. (1989) *Journal of Composite Materials*, 23, 370–88.

Quinn, J.A. (1977) *Design Data*, Fibreglass Ltd.

Rojstaczer, S., Cohn, D. and Marom, G. (1985) *Journal of Materials Science Letters*, **4**, 1233–6.

Strife, S.R. and Prewo, K.M. (1979) *Journal of Composite Materials*, 13, 264–77.

M. Orpin at BP Chemicals.

Trade data.

7.1 THERMAL PROPERTIES

Expansion

Table 7.3 Thermal expansion properties of composites

material	matrix	V_f (v/o)	CTE_l ($\times 10^{-6}$ °C^{-1})	CTE_t ($\times 10^{-6}$ °C^{-1})
carbon fibre				
T300 (Toray)	3602* (Hercules)	60	0.4	50
T300 (Toray)	epoxy	60	0.3	36.5
T400H (Toray)	epoxy	60	0.4	36.1
T800H (Toray)	epoxy	60	0.1	35.5
M40 (Toray)	epoxy	60	0	33.7
M46J (Toray)	epoxy	60	−0.7	36.5
M50J (Toray)	epoxy	60	−0.7	35.5
XA (Courtaulds)	epoxy	60	0.31	31
HM (Courtaulds)	epoxy	60	−0.26	34
HTA (Akzo)	epoxy	60	−0.1	
T300 (Toray)	5208* (Narmco)	68	−0.11	25.2
T300 (Toray)	934* (Fiberite)	57	0	29
P75 (Amoco)	934* (Fiberite)	48	−1.05	34.5
P75 (Amoco)	934* (Fiberite)	65	−1.08	31.7
P75 (Amoco)	CE339 (rubber toughened epoxy)	54	−1.02	47.4
AS4 (Hercules)	APC2	62	0.24	30
IM7 (Hercules)	APC2	62	−0.16	
P75 (Amoco)	APC2	55	−0.97	
P100 (Amoco)	APC2	45	−0.96	
aramid fibre				
Kevlar H_m (du Pont)	828 (Shell)	60	0.37	
Kevlar 49 (du Pont)	828 (Shell)	60	−2.33	
Kevlar 49 (du Pont)	PR286 (3M)	50	−2.1	68.6
Kevlar 49 (du Pont)	MY750 (Ciba Geigy)	60		76
glass fibre				
500/525 pultrusion	polyester (unspecified isophthalic)		9.4	
625 pultrusion	vinyl ester		9.4	
E (unspecified)	epoxy	40–75	4–11	
E (unspecified)	polyester	30–60	5–14	
SMC – Flomat (Freeman Chemicals)	polyester	18	20	
DMC – Beetle (BIP)	polyester	12	5	
E-glass CSM	phenolic (acid cure)	22	15	

* Modified high performance epoxy.

■ 7.1 THERMAL PROPERTIES

■ **Conductivity**

Thermal conductivity

Table 7.4 details the thermal conductivity of some fibre composites.

Materials

The manufacturers of the various fibres and matrices are given in Table 7.4. Note that Courtaulds have ceased making carbon fibres but information on their material has been included here since other manufacturers produce fibres with similar properties, see Figure 3.1, and the thermal properties of Courtaulds' materials may be relevant to these other fibres.

Observations

Again the extreme anisotropy for the carbon fibre composites is clear with the highest values of CTCs being for the very high modulus fibre materials. All the measurements recorded here were made in the vicinity

Table 7.4 Thermal conductivity data

material	matrix	V_f (v/o)	CTC_l (W/m.K)	CTC_t (W/m.K)
carbon fibre				
T300 (Toray)	epoxy	60	3	0.59
T400H (Toray)	epoxy	60	5	0.71
T800H (Toray)	epoxy	60	6	0.59
M40J (Toray)	epoxy	60	24	1.0
M46J (Toray)	epoxy	60	26	1.1
M50J (Toray)	epoxy	60	38	1.1
M55J (Toray)	epoxy	60	54	1.1
M50 (Toray)	epoxy	60	54	0.96
XA (Courtaulds)	epoxy	60	11.3	0.67
HM (Courtaulds)	epoxy	60	51	0.96
HTA (Akzo)	epoxy	60	17	
P75 (Amoco)	epoxy	60	114	1.2
aramid fibre				
Kevlar 49 (du Pont)	DX210 (Shell)	64.4	1.25	
Kevlar 49	DX210 (Shell)	57.8–67.6		0.405–0.41
glass fibre				
E	epoxy	40–75	0.3–0.35	
E	polyester	30–60	0.3–0.35	
E	epoxy		0.55–0.65	0.5
S	epoxy		0.85–1.0	
E	SC 1008 phenolic (Monsanto)	45		0.49–0.6
SMC	polyester	18	0.16–0.26	
DMC (Owens Corning Fiberglas)	polyester	12	0.28	
E-glass	phenolic (acid cure)	22	0.22	

of room temperature. The longitudinal CTC increases with temperature to at least room temperature and possibly up to approaching 100°C. The longitudinal conductivity of carbon fibre composites is proportional to the fibre volume loading.

Design implications

The anisotropy in the conductivity of carbon fibre composites can be both an advantage and disadvantage. Furthermore, the very high longitudinal CTC associated with very high modulus fibre composites could be usefully exploited in thermal designs.

Sources

Hancox, N.L. (1987) AERE-R12581.

Harris, J.P., Yates, B., Batchelor, J. *et al.* (1982) *Journal of Materials Science Letters*, **17**, 2925–31.

Mayer, R.M. (ed.) (1993) *Design with Reinforced Plastics*, Design Council, London.

Mottram, J.T. and Taylor, R. (1987) *Composite Science and Technology*, **29**, 211–32.

Quinn, J.A. (1977) *Design Data*, Fibreglass Ltd.

M. Orpin at BP Chemicals.

Trade data.

■ 7.1 THERMAL PROPERTIES

■ **Specific heat**

Specific heat

Table 7.5 gives the specific heats of various fibres.

Materials

The manufacturers of the various fibres are given in Table 7.5.

Observations

The results are reported for fibres. The measurement for Kevlar 49 was made at 50°C, the others presumably at room temperature. The specific heat increases with rising temperature. For Kevlar 49 C_p = 1220 J/kg.K at 0°C rising to 2620 J/kg.K at 200°C. A similar trend has been noted for several types of carbon fibre.

Table 7.5 Specific heat of reinforcing fibres

material	specific heat C_p (J/kg.K)
carbon fibre	
T300 (Toray)	710
T400H (Toray)	710
T800H (Toray)	710
M40J (Toray)	710
M46J (Toray)	710
M30 (Toray)	710
M40 (Toray)	710
M46 (Toray)	710
M50 (Toray)	710
P75 (Amoco)	1000
P100 (Amoco)	1000
P120 (Amoco)	1000
aramid fibre	
Kevlar 49 (du Pont)	1600
Twaron HM (Akzo)	1420
Technora (Teijin)	1090
glass fibre	
E (Vetrotex)	840
R (Vetrotex)	840
D (Vetrotex)	836
ceramic fibre	
Tyranno SiC (Ube Industries)	790

Sources

Trade data.

7.2 ELECTRICAL PROPERTIES

The electrical properties of composites that are of interest include the relative permittivity or dielectric constant, ε', the dissipation factor, tan δ, the dielectric strength (kV/mm) and the surface and volume resistivities. These properties will depend upon the constitution of the composite, the alignment of the fibres, the temperature and other enviromental conditions, and upon the frequency at which the measurement is made.

There is a certain amount of information available for the fibres and matrices and this is also included here.

■ 7.2 ELECTRICAL PROPERTIES

■ **Fibres**

Fibre properties

Some electrical properties of fibres are given in Table 7.6.

Materials

The manufacturers of the various fibres are given in Table 7.6.

Observations

The differences in resistivity among the various types of fibre are very great. Particularly noticeable is the range of resistivities provided by the

Table 7.6 Electrical properties of fibres

material	ε′	frequency	tan δ	frequency	resistivity (Ω.m)
carbon fibre					
T300 (Toray)					20×10^{-6}
T400 (Toray)					16×10^{-6}
T800 (Toray)					14×10^{-6}
T1000 (Toray)					14×10^{-6}
M35J (Toray)					11×10^{-6}
M40J (Toray)					10×10^{-6}
M46J (Toray)					9×10^{-6}
HTA (Akzo)					15×10^{-6}
STA (Akzo)		information on carbon fibres is not available			15×10^{-6}
HM45 (Akzo)					10×10^{-6}
XA (Courtaulds)					14×10^{-6}
HM (Courtaulds)					8.7×10^{-6}
P75 (Amoco)					4×10^{-6}
P100 (Amoco)					1.3×10^{-6}
P120 (Amoco)					1×10^{-6}
aramid fibre					
Kevlar 49 (du Pont)	3.8–4		0.01		5×10^{13}
glass fibre					
E	6–6.4	1 MHz	3×10^{-4}	1 MHz	4×10^{14}
S (Owens Corning Fiberglas)	5.2		6.8×10^{-3}		9×10^{12}
R (Vetrotex)	6.2	1 MHz	1.5×10^{-3}	1 MHz	
D (Vetrotex)	3.85		5×10^{-4}		
ceramic fibre					
polycrystalline metal oxide					
Nextel 312 (3M)	5.2	9.375 GHz			
Nextel 440 (3M)	5.7	9.375 GHz	0.015	9.375 GHz	
Nicalon SiC (Nippon Carbon)	4–11		0.03		$1–10^{5}$
Tyranno SiC (Ube Industries)	5–20	10 GHz	0.35		$10^{-2}–10^{5}$

SiC fibres, which may be useful in designing radar absorbing materials. The different values are obtained by varying the heat treatment temperature of the fibre.

Sources
Trade data.

■ 7.2 ELECTRICAL PROPERTIES

■ **Resins**

Resin properties

Some electrical properties of resins are given in Table 7.7.

Materials

There are many varieties of polyesters and epoxies. Neither phenolics nor bismaleimides are ideal for casting as unreinforced resin and this may account for the relative lack of properties. There are various sources of each material, but PES and PEEK are produced by ICI.

Observations

The dielectric constants are similar for all the matrices, as are the loss tangents. The resistivities are large, comparable with those of aramid and glass fibres.

Table 7.7 Electrical properties of resins

material	ε′	frequency	tan δ	frequency	resistivity (Ω.m)	dielectric strength (kV/mm)
polyester	3.2–3.7	50 Hz	$3.2–8 \times 10^{-3}$	50 Hz	7×10^{16}	12
	3.1	10 GHz	8×10^{-2}	10 GHz		
phenolic (hot cure)			5×10^{-2}	1 MHz	3×10^{9}	5.6
epoxy	3.4–6.8	50 Hz	$4 \times 10^{-3}–3 \times 10^{-2}$	50 Hz	$10^{13}–10^{15}$	9–13
	3.1–5.7	1 MHz	$1.5–6 \times 10^{-2}$	1 MHz		
bismaleimide	3.1–4.1					
PES	3.5	60 Hz	10^{-3}	1 Hz	$10^{15}–10^{16}$	
	3.5	1 GHz	3.5×10^{-3}	1 GHz		
PEEK	3.2	50 Hz	3×10^{-3}	1 Hz	$4–9 \times 10^{16}$	

Sources

Trade data.

Composite properties

Some electrical properties of composites are given in Table 7.8.

Materials

XA and HM carbon fibres are Courtaulds products; note that Courtaulds have ceased making carbon fibres but information on their material has been included here since other manufacturers produce fibres with similar properties, see Figure 3.1, and the thermal properties of Courtaulds' materials may be relevant to these other fibres. Aramid fibres used were Kevlar 29, Kevlar 49, Kevlar H_m (du Pont). E-, R- and D-glass fibres were used and Tyranno SiC and Nicalon SiC ceramic fibres (Ube Industries and Nippon Carbon, respectively). 934 (Fiberite) and MY720/HT976 (Ciba Geigy) epoxy resins were used (among others), as was polyester resin and acid cure phenolic resin.

Observations

The difference between insulators (e.g. aramid and glass composites) and conductors (carbon composites) is marked. In addition the anisotropy of resistivity is high for carbon composites. The results for Nicalon SiC were included to show the effect of systematically varying the resistivity of the fibre. The acid cure glass phenolic system has a relatively low resistivity and its tracking and arc resistance are poor compared with those of the phenolic prepreg system.

Design implications

If an electrical insulator is required then a composite based on aramid or glass fibre should be used. Carbon fibre composites can be used for electromagnetic screening.

Sources

Crone, G.A.E., Rudge, A.W. and Taylor, G.W. (1981) *IEE Proc.*, **128**, part F, December 1981, 451–64.
Quinn, J.A. (1977) *Design Data*, Fibreglass Ltd.
BIP Chemicals Ltd.
BP Chemicals Ltd.
Electra Isola.
Freeman Chemicals Ltd.
Trade data.

■ 7.2 ELECTRICAL PROPERTIES

■ **Composites**

Table 7.8 Electrical properties of composites

material	V_f (v/o)	ε′	frequency	ε′ tan δ	tan δ	frequency	resistivity (Ω.m)	dielectric strength (kV/mm)
XA carbon/epoxy longitudinal	60						50×10^{-6}	
transverse							7500×10^{-6}	
HM carbon/epoxy longitudinal	60						30×10^{-6}	
transverse							5000×10^{-6}	
Kevlar 49/polyester	44.7	3.28	9.45 GHz		0.01	9.45 GHz		
epoxy	47.1	4.43	9.45 GHz		0.01	9.45 GHz		
Kevlar H_m satin weave/Fiberite 934 epoxy		4.14	1 kHz		0.01	1 kHz	5.7×10^{13}	
		3.9	1 MHz		0.014			
Kevlar 29 satin weave/Fiberite 934 epoxy		4.19	1 MHz		0.017	1 MHz		
E-glass/epoxy	46			0.11		1 kHz		
E-glass/epoxy				0.11		10 GHz		9.8
R-glass/epoxy	46			0.15		1 kHz		
				0.089		10 GHz		
D-glass/epoxy	46			0.085		1 kHz		
				0.06		1 MHz		
Nicalon SiC plain weave/epoxy		9.76	10 GHz		0.35	10 GHz	150	
		9.16	10 GHz		0.3	10 GHz	350	
		8.46	10 GHz		0.39	10 GHz	480	
		5.6	10 GHz		0.1	10 GHz	2500	
		3.9	10.43 GHz		0.33	10.43 GHz	5×10^{5}	
Tyranno SiC/MY720 HT976 epoxy (Ciba-Geigy)	50	5–18	10 GHz					
SMC/polyester and DMC/polyester	various	4.4–4.6	1 MHz		0.015–0.02	1 MHz	$10^{14}–10^{15}$	10–15
DMC/polyester moulding compound	various	4.1–5.3	1 MHz		0.008–0.017	1 MHz		1.5–11.7
Prepreg phenolic/E-glass fabric	50	5	1 MHz		0.04	1 MHz		7
Acid cure phenolic/E-glass fabric		8	1 MHz		0.05	1 MHz	3×10^{7}	5.6

7.3 DESIGN STRATEGY

It is possible to get a wide range of thermal and electrical properties from composites. The longitudinal thermal expansion of unidirectional composite is governed by that of the fibre and is small, possibly negative, for carbon fibre composites, while in the transverse direction expansion is much larger as the contributions from both constituents are usually large in this direction. Thermal conductivity can be very high in the longitudinal direction with certain types of carbon fibre, but for other reinforcements (glass, aramid) is low. Transversely the conductivity of a unidirectional composite is small irrespective of the reinforcement.

Reinforcements, with the exception of carbon and SiC fibres, are electrical insulators, as are all polymer matrices whether thermoset or thermoplastic. Hence composites based on glass or aramid fibres can be used as insulators or in situations where transparency to electromagnetic radiation is required (e.g. D-glass). Materials based on carbon fibres may be used for electrical screening.

7.4 REFERENCE INFORMATION

ASTM D 648-72 (reapproved 1978) – measurement of HDT.

Engineered Materials Handbook, Volume 2: Engineering Plastics (1987) ASM International, Metals Park, Ohio.

Young, R.M. (1981) *Introduction to Polymers*, Chapman & Hall, London.

8 FIRE PERFORMANCE

SUMMARY

The performance of fibre-reinforced plastics is assessed in relation to the various phases in the development of a fire. Tests relating to the phases are described and material response discussed. A strategy is outlined which requires both attention to the design and the material choice. This may well limit the properties available to the designer.

8.1 INTRODUCTION

The different phases in the development of a fire are shown in Figure 8.1, and discussed in more detail in Mayer (1993).

The consequences of such a fire will include:

- heat (direct, radiant);
- flame;
- smoke obscuration;
- emission of toxic gases;
- irritation to eyes, lungs and other organs.

Human tolerance to these characteristics varies widely and so the rate of build-up is an important consideration in designing a suitable strategy against fire (section 8.8).

A major difference between thermosets and thermoplastics is that the latter may drip, causing burns or fires on anything underneath.

Relevance of data in literature

The data describe measurements of the response of materials to the types of flame or radiant heat flux. These need however to be related to the usage of the material in a component or structure.

A series of tests is required to characterize the reaction of products to different fire situations. Such tests are most useful when a range of ignition sources and heating conditions can be used so results based on a restricted range of tests should be used with caution. For example, a product may react entirely differently when exposed to a high heat flux than it does when tested with a low heat flux.

Fire retardant additives are added to resins by the suppliers to improve their fire performance. In selecting such resins, care must be taken that the additives in turn do not emit toxic species or have other undesirable effects.

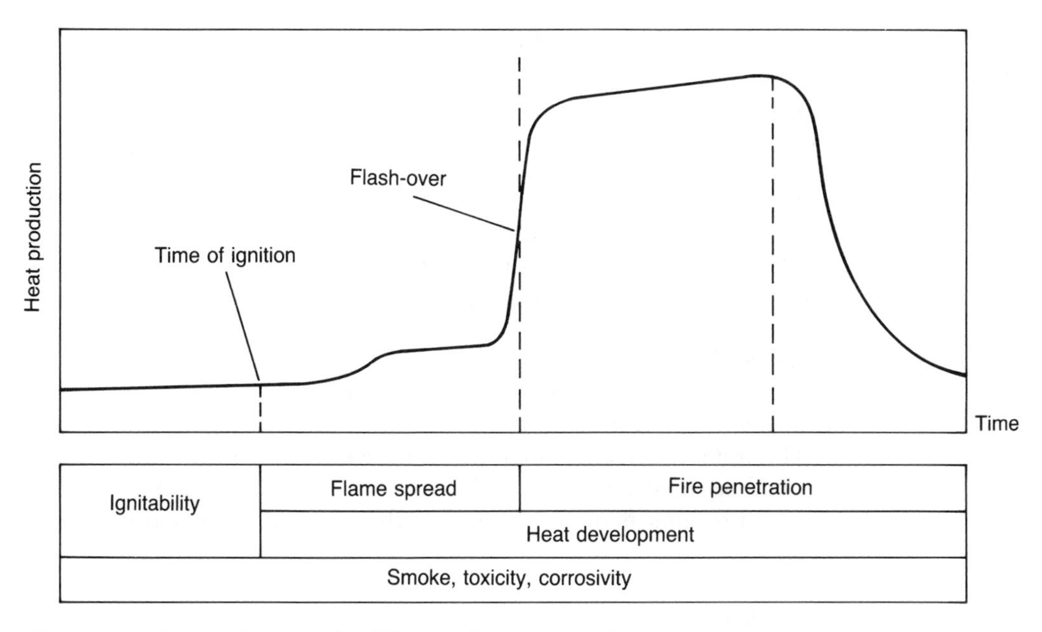

Figure 8.1 Diagram showing the different phases in the development of a fire within an enclosed space (ISO/TR 3814).

Relevance of test methods

The single market within Europe is the driving force in the transition from national to international tests. However, different tests are still required in various countries and industries at present (1992) and the designer will have to check what is permissible. Similarly, test requirements vary from one industry to another.

■ 8.2 NON-COMBUSTIBILITY

8.2 NON-COMBUSTIBILITY

Test requirements

BS 476, part 4 requires a test specimen to be inserted into the furnace at a furnace temperature of 750°C.

The material is deemed non-combustible if during the test none of the specimens cause the temperature reading, from either of two adjacent thermocouples, to rise by more than 50°C nor if any specimen is observed to flame continuously for 10 seconds or more inside the furnace.

Applicability

Experience has shown that materials meeting this requirement will not contribute significantly to a fire.

This test is intended for materials used in the construction and finishing of buildings or structures. It is also specified where fire resistant elements are used to limit flame penetration such as in bulkheads and partitions in ships.

Performance

This is a very tough test for any reinforced plastic because of the exposure to high intensity heat well above the melting point and ignition point of the resin. It is notionally impossible for materials with greater than 2% organic content to survive this test in its present form.

If the material forms part of a structure, it must not contribute to the early collapse of the structure.

Other tests

ISO 1182.
SOLAS IMO A 472.

8.3 IGNITABILITY

Test requirements

ISO 5657 measures the ignitability of a horizontally mounted essentially flat product when its upper surface is exposed to thermal radiation in the presence of a small pilot flame (Figure 8.2).

The sample is exposed to thermal radiances of up to 50 kW/m^2 from a cone-shaped heating element. A pilot flame is applied every 4 seconds to a position close to the centre of the upper surface of the test sample and remains in position for 1 second.

If sustained surface ignition occurs then the test is immediately stopped and the time noted. The test is also discontinued after 15 minutes if sustained surface ignition has not occurred. If ignition does not occur with any of 5 samples at one irradiance, then it is unnecessary to test at lower irradiances.

Applicability

The test provides systematic data on a wide range of building products and other materials under a range of intensities. It may be unsuitable for thermoplastic materials, which soften under irradiation and draw away from the pilot flame. The minimum time acceptable for ignition will depend upon the application.

Performance

Materials vary widely in this test. If the materials do not ignite under these conditions, the rate of flame spread and the possibility of flash-over will be reduced.

Other tests

BS 476 part 12 (1991) (sample exposed to a flame source rather than radiation; flame size to be appropriate to perceived hazard).

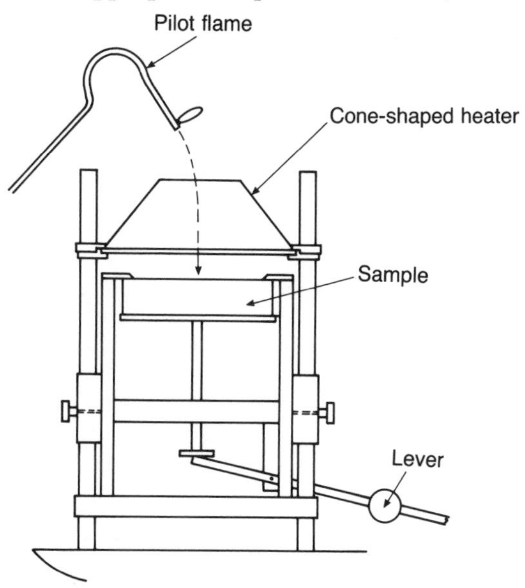

Figure 8.2 Apparatus for testing ignitability of materials (ISO 5657).

■ 8.4 FLAME SPREAD

■ **UL 94 test**

Oxygen index

8.4 FLAME SPREAD

There are a multiplicity of tests, which cover various aspects of determining the rate of flame advance. UL 94 and oxygen index tests are widely used (and quoted) to characterize materials.

UL 94 test

The most quoted test for comparing specific materials is UL 94, which assigns a flame class based on the burning behaviour at a specified thickness when ignited with a small ignition source (laboratory burner).

The lowest flame class is HB, which restricts the rate at which a flame can advance along a horizontal test specimen to 75 mm/min (<3.2 mm thick) and 38 mm/min for thicker samples. There are four classes of vertical test samples V-0, V-1, V-2 and 5V, of which 5V is the most demanding in terms of its ability to self-extinguish.

Oxygen index tests

These tests measure the minimum oxygen concentration, in an admixture with nitrogen, required to sustain combustion of small samples under specified test conditions. The higher the index, the more difficult it is for the flame to spread.

8.4 FLAME SPREAD

Material selection

Fibre and resin type

Effect of resin and fibre type

Some typical data for flame spread are listed for resins in Table 8.1, and for fibres in Table 8.2.

Observations

Both resins and fibres differ widely in their performance under flame conditions.

For polyester resins, three major possibilities exist for improving fire performance:

- inert fillers like antimony trioxide with chlorinated hydrocarbons;
- incorporation of halogenated species in the polymer backbone;
- water-releasing fillers like aluminium trihydrate.

For thermoplastic resins, one needs to select a resin type and blend with a suitable set of fire and mechanical properties in discussion with the manufacturer, who does the formulation.

Design implications

If the various constituents are to be used in a composite then it is the performance of the composite that matters. For example, the oxygen

Table 8.1 Limiting oxygen index and UL 94 rating of some resins

		UL 94 rating		
material	oxygen index (%)	best	average	worst
epoxy	20–30	V-0	range	HB
PA 6/6	21–32	V-0	range	HB
phenolic	25–36	V-0	range	HB
polyester	20–80	V-0	V-0	HB
PES	34–40	V-0		
PPO	17–28		V-0	HB

Table 8.2 Flammability of principal fibres

fibre	performance	comment
glass	inorganic, does not burn	softens at 750°C flows *c*.1200°C
aramid	oxygen index 27–29% vertical flammability test	 no drips no after-burn time glow time increases and burn length decreases with increasing fabric mass
carbon	burns *c*.600°C	carbon dioxide given off

■ 8.4 FLAME SPREAD

■ **Material selection**

Fibre and resin type

index for glass mat/phenolic (Table 8.4) is much greater than that of the resin alone because of the screening effect of the glass, which does not burn.

Other tests

- oxygen index BS 2782 part 1 method 141;
  Nordtest NT FIR 013;
  ASTM D 2863.

Sources

Waterman, N. and Ashby, M. (1991) *Elsevier Materials Selector*, Elsevier, Barking.
du Pont trade data.
ASM (1987) *ASM Handbook Engineering Plastics*, Metals Park, Ohio.

Surface spread

ISO/DIS 5658 part 1 test is used to measure the rate of travel of a flame front in a horizontal direction over the surface of an essentially flat specimen exposed to thermal radiation (Figure 8.3).

The specimen (800 × 155 mm) is positioned so that its surface is exposed to thermal radiation from a vertical mounted gas-fired radiator, the irradiance decreasing along the specimen's length. A pilot flame is used to initiate flaming at the high radiance end of the sample.

As a flame front develops, its lateral progress is assessed and the time to initiate and extinguish and the maximum distance travelled are recorded. The test is continued for a minimum period of 10 minutes.

Applicability

The test is applicable to many products and is under consideration for characterizing certain classes of building products.

Other tests

- flame spread BS 476 part 7 (British);
  NEN 3883 (Dutch);
  DIN 4102 parts 1, 15, 16 (German; burner size to be specified);
  NFP 92501 (French; epiradiateur);
  IMO Res. A653 (16) (adopted 1989, heat release as well as flame spread);
  ASTM E 84 (American; Steiner tunnel test; flame spread and smoke).

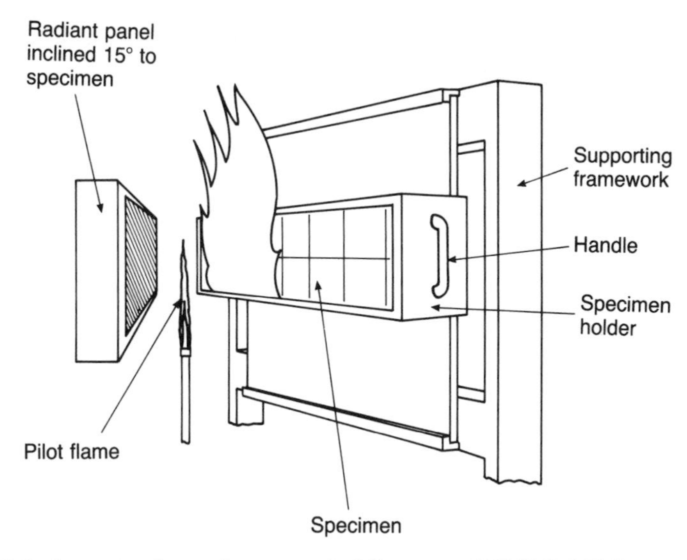

Figure 8.3 Apparatus for surface spread of flame test (ISO/DIS 5658).

Fire resistance

The method is to assess the behaviour of a specimen representative of an element of a building construction under conditions applicable in practice. Wherever possible, full size elements are to be tested with a minimum size of 3 m.

Fire resistance is expressed as the time during which the appropriate criteria of load bearing capacity, fire containment (integrity) or thermal insulation functions can be maintained. An example is given in Table 8.3 for a glass/phenolic laminate where a mean maximum temperature rise of 140°C is allowed above the initial mean temperature on the unexposed face.

Other standards

ISO 834, ISO 3008, ISO 3009, EUR 8750.

Table 8.3 Time for temperature to increase on an unexposed face for a 9 mm thick glass/phenolic laminate exposed to a representative fire source

temperature (°C)	(V_f = 49 v/o) min	(V_f = 53 v/o) min
50	3	3
100	4	4
160	19	26
200	22	28

Sources

M. Orpin at BP Chemicals.

BS 476 part 20, a method for the determination of the fire-resistant elements of construction.

Influence of resin type

Phenolic

Some data for the flammability of glass-reinforced phenolic resin are given in Table 8.4.

Methacrylate

Aluminium trihydrate (ATH) can be added to a methacrylate resin to improve its fire resistance (Figure 8.4).

Note that the glass fraction has a significant effect on the fire performance, as the higher the fraction, the lower the amount of filler required to obtain a given rate of surface spread.

Some flammability data are listed in Table 8.5 for a commercially formulated dispersion of aluminium trihydrate in a glass-reinforced methacrylate resin.

Materials

- resin – methacrylate 835 S with aluminium trihydrate (ICI);
- glass – 20 w/o CSM.

Observations

Phenolic has inherently good fire performance due to its ability to form a char on the surface in contact with the flame, thus cutting off the oxygen supply to the resin beneath.

Aluminium trihydrate is an effective fire retardant agent and has been successfully incorporated into methacrylate as well as polyester resins.

Table 8.4 Flammability properties of phenolic composite

test	method	classification
fire propagation	Building Regs B2 parts 3 and 4, BS 476 part 6	Class 0 $i < 6$; $I < 12$
surface spread of flame	BS 476 part 7	Class 1
	DIN 4102	B1
	NEN 3883	Class 1
	NFF 16.101	M1/F1
oxygen index	ASTM D 2863	45–80%

Table 8.5 Flammability properties of a methacrylate resin containing aluminium trihydrate as filler

fire test	classification
DIN 4102 (Brandschacht)	B1
BS 476 part 7	1
NFP 92501 (Epiradiateur)	M2
UL 94	V-0

8.4 FLAME SPREAD

Glass fabric

Phenolic and methacrylate

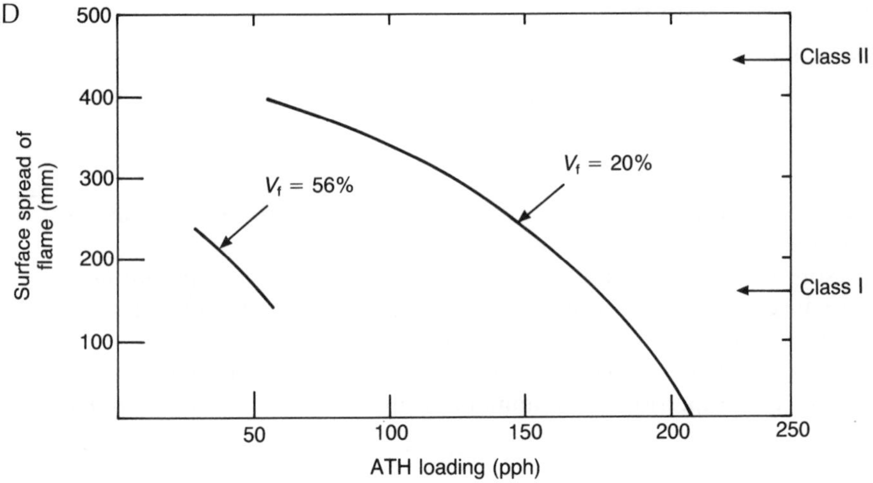

Figure 8.4 Surface spread of flame for methacrylate composites as a function of aluminium trihydrate loading; measurements using BS 476, part 7. Arrows indicate flammability classes (Sayers).

The glass content also has a significant effect on fire performance with higher volume fractions providing better resistance to flame spread.

Design implications

Good resistance to flame spread can be secured either by choosing a resin which is inherently resistant or by adding a suitable filler to a thermosetting resin.

A variety of tests is currently (1992) necessary to determine flame spread, but this will change as test methods are agreed at international level.

Sources

Sayers, D.R. *et al.* (1988) Development of low smoke, zero halogen, fire retardant composites based on methacrylate resins, *Proc. British Plastics Federation Conference*, London.

M. Orpin at BP Chemicals.

8.5 RATE OF HEAT RELEASE

Test requirements

ISO 5660 describes a method for determining the rate at which heat is evolved from a fire using an oxygen consumption calorimeter. It is based on the principle that for a range of materials, the heat released is proportional to the amount of oxygen required for combustion.

A 100 mm square sample is exposed to irradiance in the range of 10 kW/m^2 to 100 kW/m^2 from a cone-shaped heater (Figure 8.5). The pyrolysis gases produced are ignited by a spark and the resulting combustion gases are extracted through an exhaust system.

Mass loss can also be determined as well as smoke obscuration.

Applicability

The test applies to products which have flat surfaces, but is not suitable for testing very thin samples or certain types of composites.

Performance

This test is emerging as an internationally agreed method of determining the response of materials to heat and flame under controlled laboratory conditions. Aerospace continues to use the OSU Rate-of-Heat Release chamber as prescribed by the airworthiness authorities.

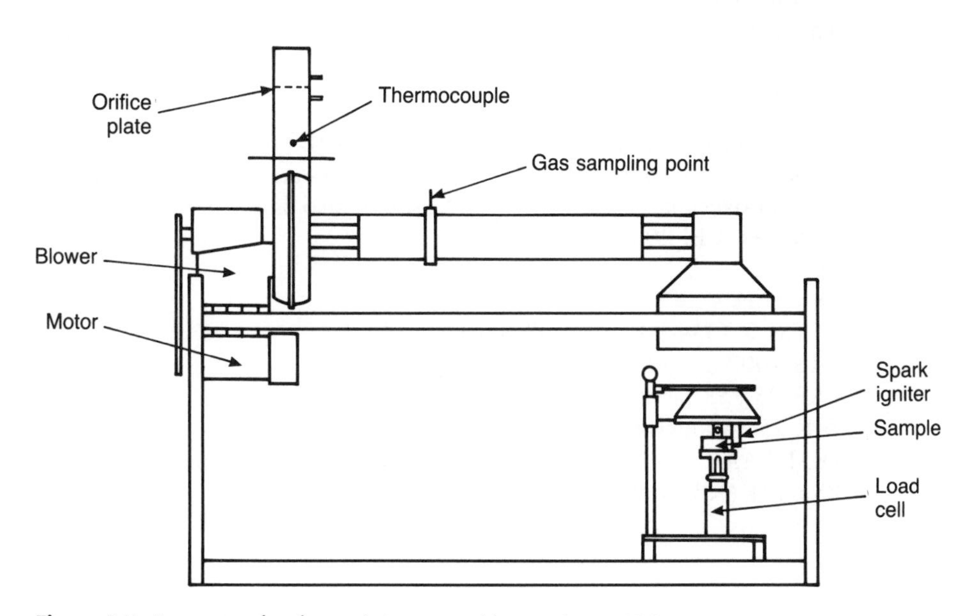

Figure 8.5 Apparatus for determining rate of heat release (ISO 5660).

Source

ISO 5560.

■ 8.6 GAS EMISSION

■ **Test methods**

8.6 SMOKE AND TOXIC GAS EMISSIONS

Smoke can be defined as a mixture of particles, liquid aerosols and gaseous products resulting from a combustion process. The toxicity of the gaseous products is of prime concern as a direct threat to life.

Smoke

Obscuration is typically measured in the NBS smoke chamber (ASTM E 662). This is a small scale laboratory test of a sample exposed to a radiant energy source, either with or without an adjacent pilot flame to ignite the gases from the material.

The attenuation of a beam of visible light is recorded against time to derive the specific optical density. The standard heat flux is 25 kW/m^2.

The ISO dual chamber test (ISO/TR 5924) requires larger samples (165 mm square), and exposure up to irradiances of 50 kW/m^2 from a cone-shaped heating element (such as shown in Figure 8.2, but without the pilot flame).

Toxicity

Testing involves identifying the nature, concentration and injurious effects of the major gaseous species. The rate of evolution of the toxic species will in turn depend upon the type of fire.

Applicability

The concern is primarily within confined spaces like buildings and transport. The need for large scale tests is apparent and this has been tackled in various ways, for example,

- underground trains – 3 m cube smoke test (BS 6853);
- buildings – room test (NT fire 025) (also rate of heat release and flash-over), large scale test (ASTM E 84) (also flame spread), electrical cables (CEI 20.22).

Other tests

ASTM E 1354-90 (cone calorimeter).
ATS 1000.001 (Airbus Industrie Spec.).
FAR 25.853 (aerospace).
JAR 25.853 (aerospace).
DOT/FAA/CT-89/15 Aircraft Material Fire Test Handbook.
BS 6401 (NBS smoke box).

8.6 GAS EMISSION ■

Industry limits ■

Performance

Some typical self-imposed, aerospace industry limits for smoke and toxic gas measurements are listed in Tables 8.6 and 8.7, but their relation to human exposure in real fire situations is not yet known.

Table 8.6 Smoke emission limits for NBS smoke chamber, 25 kW/m^2, flaming and non-flaming modes

material	specific optical density (limits within 4 min)
ceiling and side wall panels	150
cargo liners	150
air ducting, thermal and acoustic insulation	100
carpets	200

Table 8.7 Typical limits for toxic gas emission in NBS smoke chamber, 25 kW/m^2, flaming and non-flaming modes

	concentrations within 4.0 min (ppm)
HCN	150
$NO + NO_2$	100
$SO_2 + H_2S$	100
HF	100
HCl	150
CO	3500

Sources

Madgwick, T. (1988) Aircraft fire safety – are we meeting the challenge? in *Fire Performance of Materials in Mass Transport*, RAPRA, Shawbury.

Airbus Industrie Specification ATS 1000-001.

■ 8.6 GAS EMISSION

■ **Chopped strand mat**

Resin type

Influence of resin

The influence of resin type on smoke emission is shown in Figure 8.6.

Materials

- resins – phenolic J2027 (BP Chemicals), polyester fire retardant to BS 476 part 7/cl 1, epoxy (non-fire retardant);
- reinforcement – chopped strand mat.

Manufacture

Contact moulding by hand lay-up.

Observations

Phenolic has a low smoke emission compared with epoxy and polyester because the phenolic resin structure is char forming whereas the other two resins are not.

Design implications

The rate of smoke generation is of considerable importance in the early stages of a fire when people are trying to escape; the maximum smoke density will hinder those who are only able to escape at a later stage.

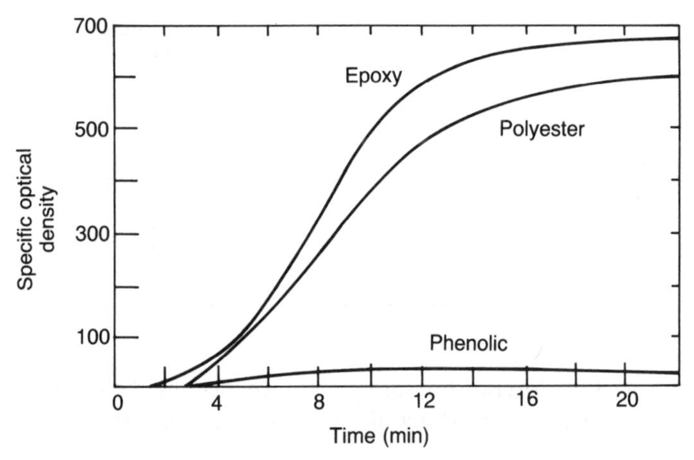

Figure 8.6 NBS smoke chamber test of CSM laminates with various resins (smouldering mode) (Forsdyke).

Source

Forsdyke, K.L. (1988) Phenolic FRP today, in *Proc. British Plastics Federation Conference*, London.

8.6 GAS EMISSION ■

Fillers ■

Chopped strand mat and resin type

Effect of fillers

The influence of fillers and resin type on smoke and toxicity is compared in Figure 8.7 and Table 8.8.

Materials

- resin – methacrylate 835 S (ICI);
- additive – 82 pph aluminium trihydrate (ATH), (or) 208 pph aluminium trihydrate (ATH);
- resin – polyester fire retardant to BS 476 part 7/c/1;
- reinforcement – CSM (20 w/o).

Manufacture

By contact moulding and hand lay-up.

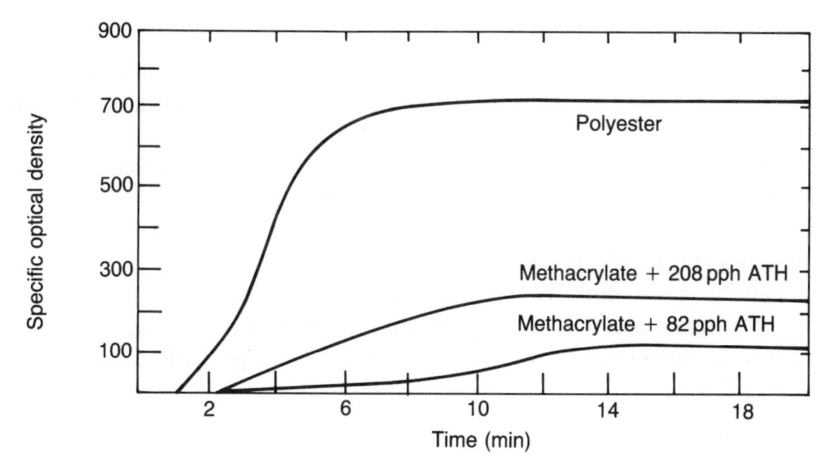

Figure 8.7 Smoke density measured by NBS smoke chamber (flaming mode) for various resins (Sayers).

Materials

- resin – methacrylate 826 HT plus 100 pph aluminium trihydrate;
- reinforcement – glass-mat/woven roving/mat combination;
- material – phenolic SMC.

Manufacture

- methacrylate composite – pultrusion;
- phenolic SMC – press moulding.

■ 8.6 GAS EMISSION

■ **Fillers**

Chopped strand mat and resin type

Table 8.8 Toxic gas emissions of reinforced methacrylate and phenolic resins compared with current New York City standard

gas evolved	methacrylate (ppm)	phenolic (ppm)	New York specification (ppm)
hydrogen halides	0	0	10 each
aldehydes as HCHO	5	4	30
ammonia	0	0	500
carbon monoxide	315	208	1000
carbon dioxide	2650	4050	10000
oxides of nitrogen	14	38	1000
hydrogen cyanide	2	<2	10

Note: Measured by MIL-M-14G test (USA).

Observations

Methacrylate resin when loaded with a suitable filler has a much reduced smoke emission compared with a conventional fire retardant polyester resin.

The filler is likewise effective in achieving low toxic emissions and the composite has comparable properties to that of phenolic SMC.

Design implications

Phenolic resin and ATH filled methacrylate resin provide good fire performance amongst the thermosetting resins. The risk assessment can only be satisfactorily undertaken if relevant large scale tests are also carried out.

Source

Sayers, D.R. *et al.* (1988) Development of low smoke, zero halogen, fire retardant composites based on methacrylate resins, in *Proc. British Plastics Federation Conference*, London.

8.6 GAS EMISSION ■

Large scale testing ■

Phenolic, polyester

Large scale tests

For underground rolling stock, London Underground have devised a large scale test, which has been incorporated into a British Standard (BS 6853). Some materials test data are listed in Table 8.9.

Manufacture
Contact moulding by hand lay-up.

Observations
Both phenolic and painted phenolic composites meet the smoke requirements. However, fire retardant polyester gel coats on phenolic resins are unlikely to meet the test. The most suitable type of paint (since phenolics cannot be gel-coated) needs to be discussed with the resin supplier.

Design implications
BS 6853 is a code of practice for fire precautions in the design and construction of railway passenger rolling stock. This provides guidance on test methods and compliance criteria, as well as design principles, the 3 m cube smoke test being just one of these requirements.

Table 8.9 Material testing for use in underground railway carriages

	3 m cube smoke test (A_0)		flammability temperature index
	smouldering	burning	
category I requirement interior (BS 6853)	<1.5	<1.0	>350°C <50 mm travel
phenolic GRP (60% resin)	0.25–0.35	0.15–0.25	>350°C nil travel
painted phenolic GRP	0.55–0.65	0.40–0.50	
typical fire retardant polyester GRP	*c.*100		
best polyester DMC (*c.*20% resin)	*c.*12		

Sources
Forsdyke, K.L. (1988) Phenolic FRP today, in *Proc. British Plastics Federation Conference*, London.
M. Orpin at BP Chemicals.

■ 8.6 GAS EMISSION

■ **Reinforcement type**

Resin type

Influence of reinforcement

Large scale tests have been undertaken by the FAA on panels inside a C130 aeroplane fuselage (Figure 8.8).

Materials
- reinforcements – glass, aramid, carbon, polyimide (PI);
- resins – epoxy, phenolic, PEEK;
- foam core – honeycomb (Nomex) dipped in phenolic resin.

Manufacture
Large test panels by prepreg moulding.

Observations
Panels reinforced with different fibres demonstrate distinct smoke profiles on ignition. Smoke initiation for phenolic resins is much slower with glass reinforcement than with graphite or aramid. The rate of generation follows a similar trend with glass being the slowest and aramid the fastest.
The variation between resin types is obvious with glass/phenolic being much better than glass/epoxy (*cf.* Figure 8.6). Polyimide/PEEK is the most resistant to smoke generation.

Design implications
There is a large variation in the build-up of smoke levels during a fire and hence choosing the right material combination will delay obscuration of vision, something of great importance in a confined space such as an aircraft.

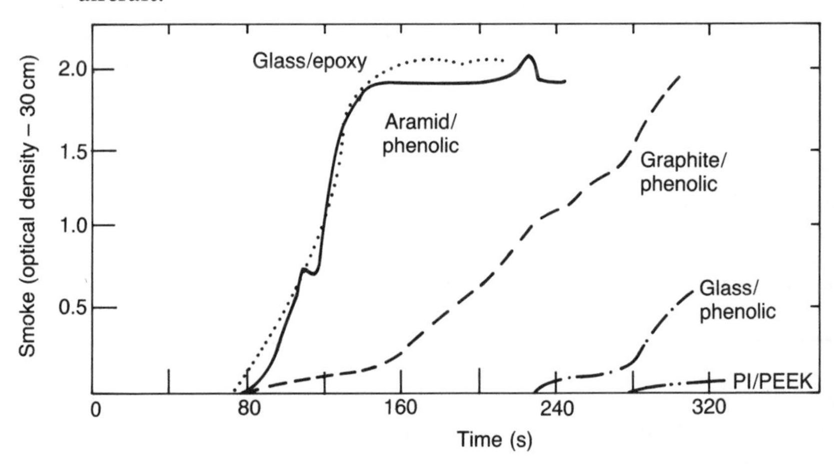

Figure 8.8 Comparative smoke profiles on panels comprising various reinforcements and resins (Madgwick/FAA).

Source
Madgwick, T. (1988) Aircraft fire safety – are we meeting the challenge? in *Fire Performance of Materials in Mass Transport*, RAPRA, Shawbury.
DOT/FAA/CT-85/23.

8.7 FLASH-OVER

Flash-over is a particular state of flame spread (Figure 8.1) which is achieved once a flammable gas concentration reaches a critical level, which causes a flash ignition. This change can cause a relatively slow burning fire to rapidly become a major conflagration, which could lead to a loss of life.

The insulation of an enclosed space has a considerable effect because heat not lost to the surroundings can cause considerable increase in the fire growth rate. Large scale tests have been carried out by the FAA on panels inside a C130 aerospace fuselage with a kerosine flame as source (Figure 8.9).

Materials

- reinforcements – glass, aramid, carbon, polyimide (PI);
- resins – epoxy, phenolic, PEEK;
- foam core – honeycomb (Nomex) dipped in phenolic resin.

Observations

Both fibre and resin affect the flash-over time with glass/epoxy being short (one minute) whilst glass/phenolics achieved four minutes. In overall trend, the flash-over time follows that of smoke generation (*cf.* Figure 8.8).

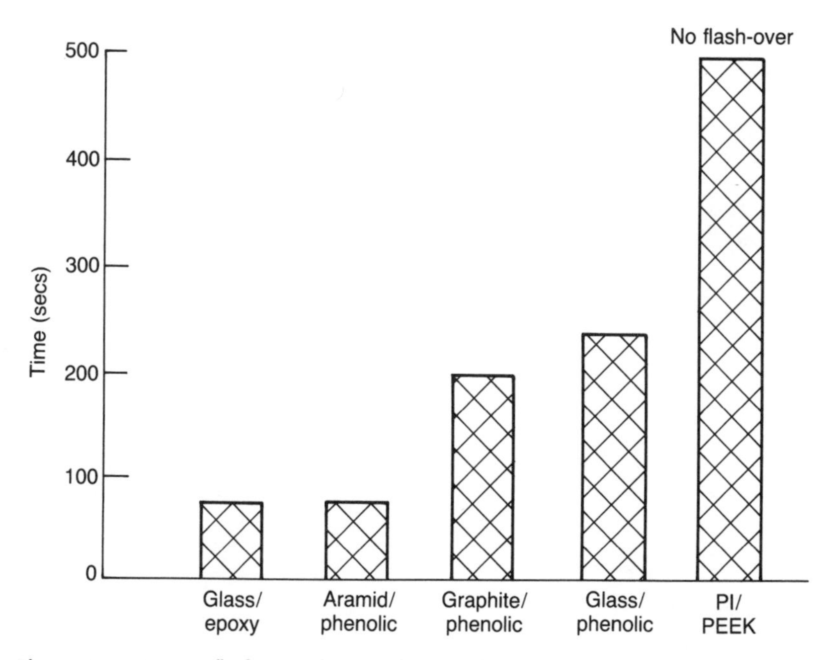

Figure 8.9 Time to flash-over for panels manufactured from a variety of reinforcements and resins (Madgwick/FAA).

■ 8.7 FLASH-OVER

■ **Reinforcement type**

Resin type

Design implications
The time needed to evacuate an aircraft will clearly set a lower limit on the flash-over time and therefore limit the material choice. Flash-over is a highly dangerous condition and can be avoided with the above fire source by choosing a thermoplastic resin/fibre system (e.g. polyimide/PEEK), or delayed by using an appropriate phenolic system (e.g. glass/phenolic).

Other tests
NEN 3883 (small scale).
Nordtest fire 025 (large scale).
ISO 9705 (large scale).

Source
Madgwick, T. (1988) Aircraft fire safety – are we meeting the challenge? in *Fire Performance of Materials in Mass Transport*, RAPRA, Shawbury.

8.8 DESIGN STRATEGY

A design strategy against fire is given in Figure 8.10.

The starting point is generally that of material selection and performance and the end point that of the consideration of structural safety (if relevant). A second aspect concerns the combustion products and what risk these pose to occupancy of a building or enclosed space like a ship, aircraft or motor vehicle.

The applicability of any of the forementioned standards should be established, and even if none is specified, it is recommended to demonstrate compliance with the nearest relevant standard.

Finally the designer needs to decide the necessary level of fire resistance, how this might be attained and the measure of fire safety required.

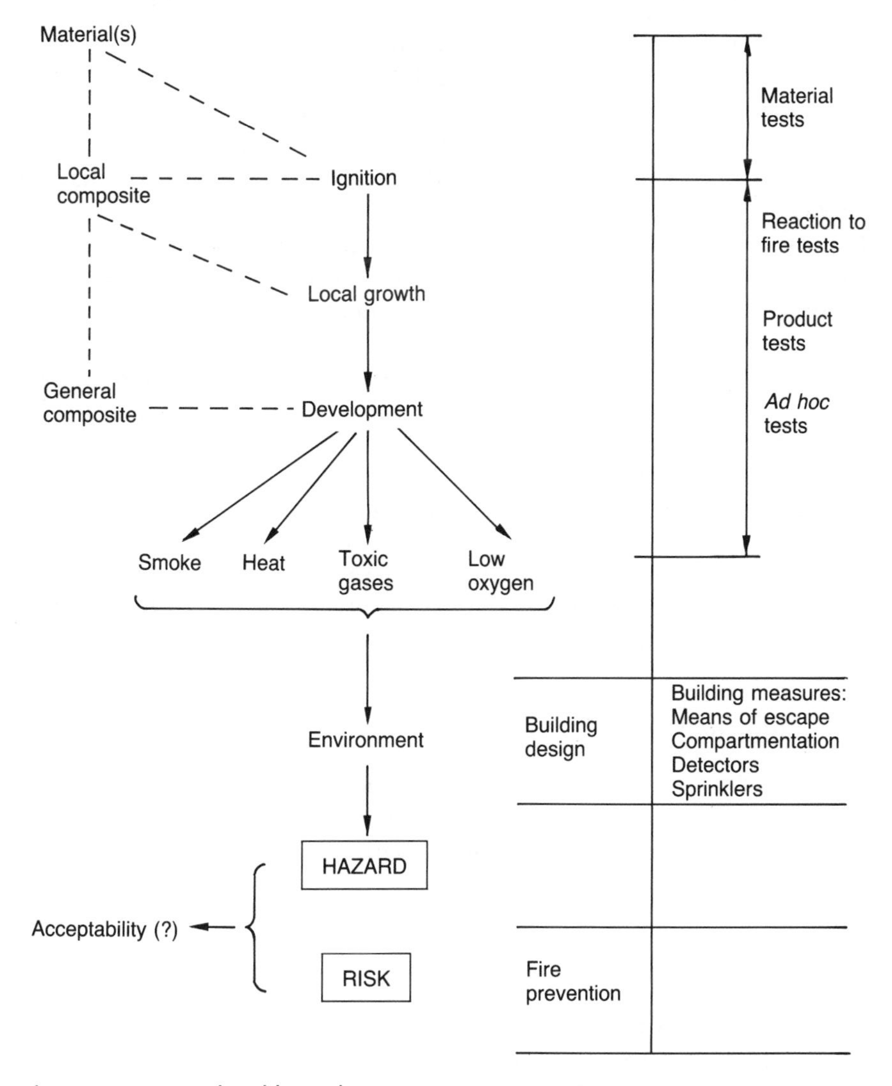

Figure 8.10 Hazard and hazard assessment (PD 6503 Part 2).

The test methodology must encompass tests on a large enough scale to ensure realistic reaction to fire tests.

As there are only generic data in this guide, additional information can be sought from:

- a database like Plascams for limiting oxygen index and UL 94 flame spread tests;
- materials suppliers for detailed information and advice.

8.9 REFERENCE INFORMATION

General references

Airbus Industrie Standard ATS 1000-001 Issue S – Appendix A – generally regarded as a definitive statement of aircraft cabin requirements.

Aircraft Material Fire Test Handbook, DOT/FAA/CT-89/15 – up-to-date presentation of procedures in use by aircraft industry.

Fire Toxicity of Plastics (1989) RAPRA Technology, Shawbury – papers presented at a seminar covering all the major aspects of fire, smoke and toxicity.

Madgwick, T. (1988) Aircraft fire safety – are we meeting the challenge? in *Fire Performance of Materials in Mass Transport*, RAPRA, Shawbury – book contains papers presented at a seminar covering all major forms of transport.

Makower, A.D. (1989) *Fire Tests – Building Products and Materials*, BSI, Milton Keynes – a standard reference providing a brief guide to the relevant British standards.

Mayer, R.M. (ed.) (1993) *Design with Reinforced Plastics*, Design Council, London, Chapter 4.9 covers designing against fire.

Extracts from British and International standards are reproduced with permission; complete copies can be obtained from national standard bodies like BSI, Linford Wood, Milton Keynes, MK14 6LE, UK, Tel. (44) 0908 22 00 22.

Non-combustibility

Rogowski, B. (1988) Fire control and material selection in ships, in *Fire Performance of Materials in Mass Transport*, RAPRA Technology, Shawbury – describes the need for isolating fires in ships and outlines the International Maritime Organization tests.

Ignitability

ISO TR 3814 (Tests for Measuring Reaction of Fire to Building Materials) (1989) ISO, Geneva – outlines tests to characterize various phases of a fire and introduces the concept of a 'tool-kit' of tests.

Flame spread

ISO/DIS 5658 – is under discussion and is likely to appear as three parts in the future.

Plascams Database (1991) RAPRA Technology, Shawbury – lists oxygen index and flame spread (UL 94) for many resins and blends, particularly thermoplastics.

Troitzsch, J. (1986) *International Plastics Flammability Handbook*, Hanser, München – principles, regulations, testing and approval.

Rate of heat release

Babrauskas, V. (1984) Development of the cone calorimeter – a bench scale heat release apparatus based on oxygen consumption. *Fire and Materials*, 8, 81–95.

Smoke and toxic gas emission

DD 180 (Guide for the Assessment of Toxic Hazards in Fire in Buildings and Transport) (1989) BSI, Milton Keynes – provides guidelines and stipulates limits for incapacitating lethal doses and concentrations of the major toxic entities in fire for use in hazard estimation.

ISO TR 9122 (Toxicity Testing of Fire Effluents) (1989) ISO, Geneva – review of current state of the art.

Flash-over

BS 476 part 20 (Method for Determination of the Fire Resistance Elements of Construction).

Design strategy

ISO TR 3814 (Tests for Measuring Reaction of Fire to Building Materials) (1989) ISO, Geneva – need for risk assessment.

PD 6503 part 2 (1988) BSI, Milton Keynes – guide to the relevance of small scale tests for measuring the toxicity of combustion products of materials and composites.

Fire Test Procedures (1984) IMO, London – describes procedures and hazards in accordance with the Safety of Life at Sea (SOLAS) convention.

Information sources

Material suppliers can advise on the selection of the most suitable resin within their range and its fire resistance properties. It is essential to verify the material combination proposed for production with the appropriate tests.

Advice can be sought in the UK from:

Building Research Station, Garston, Watford, WD2 7JR, Tel. (44) 0923 664664.

Queen Mary's College, Mile End Road, London, E1 4NS, Tel. (44) 071 980 4811.

SGS Yarsley Laboratories, Trowers Way, Redhill, Surrey, RH1 2JN, Tel. (44) 0737 765070.

Warrington Fire Research Centre (London) Ltd, 101 Marshgate Lane, London, E15 2NQ, Tel. (44) 081 519 8297.

9 ENVIRONMENTAL EFFECTS

SUMMARY

The way in which the environment can attack the resin and the reinforcement is discussed in broad outline. The effects of moisture, corrosion, abrasion, wear and weathering on properties are considered in turn. There is considerable variability amongst materials so it is possible to design components which can successfully withstand their operating environment.

9.1 INTRODUCTION

The environment in the form of a gas, liquid or particles can attack the resin and the reinforcement in various ways. Possible ways include:

- lowering the glass transition temperature of the resin by absorbing water;
- penetrating the resin;
- damaging the interface between the fibres and the resin thereby hindering load transfer from the matrix to the fibres and vice versa;
- attacking the fibres;
- abrading the surface thereby exposing the fibres.

As the reinforcement's role is to boost the resin properties, it is important that the surface be resin rich to protect the fibres from the influence of the environment.

Most data are derived for a single influence and the effect of combining more than one influence has then to be determined, e.g. combining temperature, loading and exposure to abrasive particles to characterize the environmental effects on a pump housing.

In addition, results of tests on small samples may not be representative of large panels because of the differing ratio of surface to volume. In practice, it is generally found that the small scale tests tend to give conservative results.

Differences compared with metals

Fibre-reinforced plastics differ from metals in various ways as regards environmental degradation:

- resins do not oxidize, but they can degrade in ultra-violet light and absorb water;
- resin provides a protective layer for the fibres, which carry the primary load;
- resins are more vulnerable to organic solvents than metals;
- resins and fibres differ widely in their ability to withstand the environment, so a choice is generally available for a particular application.

It is for these reasons that fibre-reinforced plastics are often preferred to metals in aggressive environments.

9.2 EFFECT OF MOISTURE

Components may be subjected to conditions of high relative humidity or directly exposed to, or immersed in, fresh or sea water and so take up moisture. In desert conditions, the converse would apply as the relative humidity would be very low and so moisture might well be given off. The rate of water uptake (or release) and its equilibrium concentration as well as its influence on properties, could therefore be of concern.

Tests in boiling water are often used as a simple way to accelerate the effect of water immersion and give an indication of the worst effects likely to occur.

■ 9.2 MOISTURE

■ **Water absorption**

Glass/polyester

Water absorption

A systematic study has been carried out on the kinetics of absorption as a consequence of prolonged immersion in water. The data for the pure resin and laminates are shown in Figures 9.1 and 9.2, together with a prediction of diffusion based on Fick's Law, which relates the diffusion rate of a species to temperature via a constant and an activation energy.

Materials

- resin – iso-polyester A283/270 (BP Chemicals);
- reinforcement – E-glass UD fabric, 2400 tex (Pilkingtons).

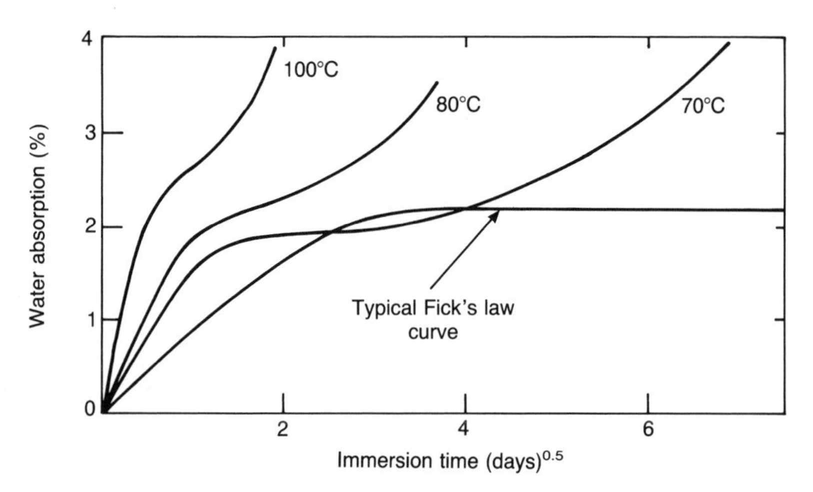

Figure 9.1 Water absorption kinetics of pure iso-polyester resin (Pritchard and Speake).

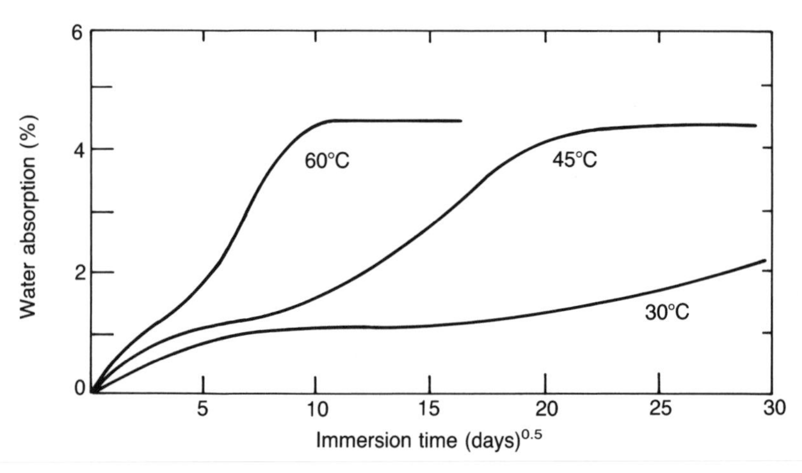

Figure 9.2 Water absorption kinetics of glass-reinforced iso-polyester resin (Pritchard and Speake).

9.2 MOISTURE ■

Water absorption ■

Glass/polyester

Table 9.1 Activation energies for diffusion of water into and out of resins and laminates

	water uptake (kJ/mol)	water desorption (kJ/mol)
iso-polyester resin	55.4	187.6
UD glass/polyester	73.4	175.8
polyester laminates	45–51	
epoxy resin	55.6	
glass/epoxy laminates	44–48	

Manufacture

- contact moulding by hand lay-up;
- fabric laid up unidirectional or balanced bi-directional (0°/90°).

Observations

The water absorption uptake of both pure resin and laminate is very dependent upon temperature. The higher the temperature the quicker the plateau value is reached. At temperatures around the glass transition temperature of the matrix (74°C), there appears to be a further uptake in water absorption possibly balancing the loss of organic material by leaching.

The diffusivity is greater in the pure than the reinforced resin. The results have been analysed in terms of Fick's Law and the activation energies for the water uptake and desorption of various materials are compared in Table 9.1.

Source

Pritchard, G. and Speake, S.D. (1987) The use of water absorption kinetic data to predict laminate property charges. *Composites*, **18**, 227–32.

9.2 MOISTURE

Mechanical properties

Glass/polyester

Effect on properties

The influence of water uptake on the tensile properties is shown in Figure 9.3, the materials being the same as described in the previous section.

Materials

- resin – iso-polyester A283/270 (BP Chemicals);
- reinforcement – E-glass UD fabric, 2400 tex (Pilkingtons).

Observations

The data, collected over a range of temperatures, lie on a single curve indicating the influence of water uptake and not the immersion temperature. Similar findings were obtained for other properties including angle ply tensile strength of the composite.

Design implications

A curve-fitting program has been developed by Pritchard and Speake to obtain empirical relationships linking various mechanical properties to absorbed water content after immersion at various temperatures for various times. The property predictions have been verified by prolonged testing at 30°C and 45°C (Figure 9.4).

Bulder and Bach discuss other results for glass-reinforced polyester. For example, the water uptake is much less (0.5% *cf.* with 4%) if the relative humidity (R.H.) is 50% rather than 100%. The change in modulus is always less than the change in strength for a given water content.

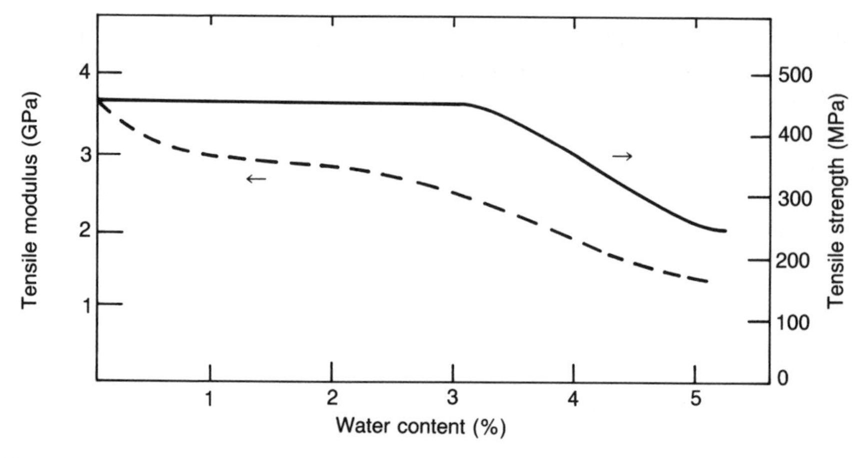

Figure 9.3 Tensile modulus of the resin; tensile strength of GRP laminates as a function of water uptake after periods of immersion at various temperatures between 30°C and 100°C up to the onset of equilibrium absorption (Pritchard and Speake).

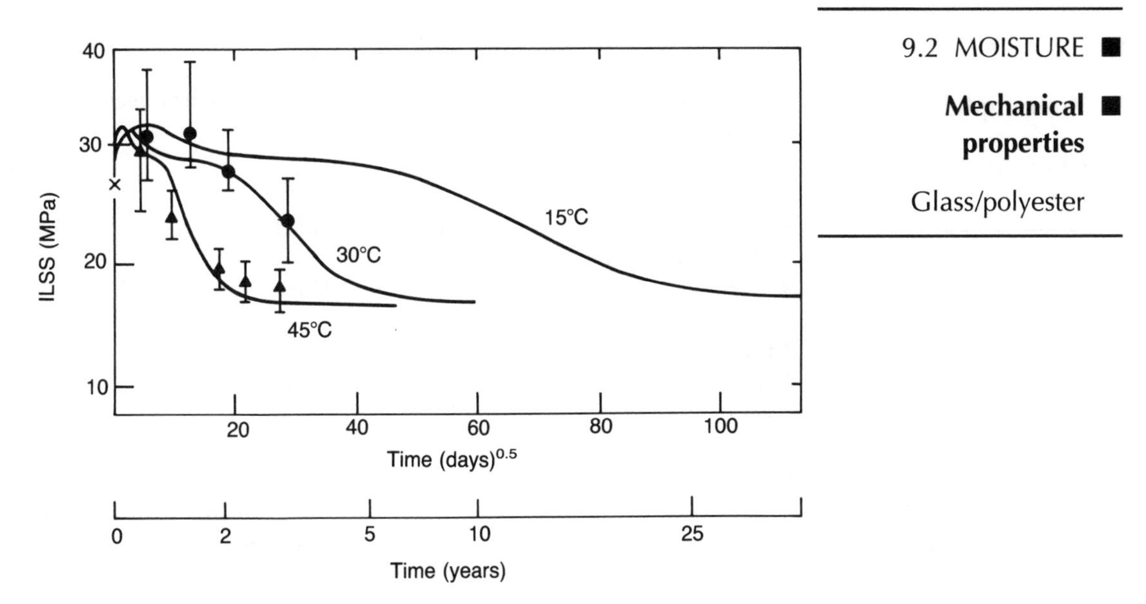

Figure 9.4 Predicted shear strength versus immersion time and comparison with experimental results (Pritchard and Speake).

Sources

Bulder, B.H. and Bach, P.W. (1991) *Literature Survey on the Effects of Moisture on the Mechanical Properties of Glass and Carbon Plastic Laminates, ECN-C-91-033*, ECN, Petten.

Pritchard, G. and Speake, S.D. (1987) The use of water absorption kinetic data to predict laminate property changes. *Composites*, **18**, 227–32.

■ 9.2 MOISTURE

■ **Water uptake**

Fibre type and epoxy resin

Effect of pre-conditioning and fibre type

The hydro-thermal conditioning of epoxy laminates has been measured for the three principal reinforcing fibres.

Materials

- reinforcement – carbon HTS (Courtaulds), aramid Kevlar 49 (du Pont), E-glass (Pilkingtons);
- resin – epoxy type 69 (Fothergill and Harvey).

Manufacture

Laminates moulded from unidirectional prepregs with balanced warp/weft.

Observations

The change in the laminates' mass describes the loss/gain of moisture and this varies from the higher water absorbency of aramid composites to lower capacities of glass and carbon composites (Figure 9.5). Saturation of a composite is achieved far more easily in boiling water than in a humid atmosphere at room temperature.

The effect of such preconditioning is to decrease the tensile properties (Table 9.2). The change is minimal for carbon and noticeable for aramid composites. Whereas glass composites are only slightly affected by a high relative humidity, boiling has a drastic effect.

Interlaminar shear strength is however increased by exposure to a humid atmosphere (Table 9.3). Boiling does not affect aramid composites whilst the glass composite is weakened by 10% and the carbon composite's shear strength is increased by *c*.20%.

Table 9.2 Effect of preconditioning treatments on the tensile properties of reinforced epoxy laminates

tensile property		treatment	glass	aramid	carbon
strength	(MPa)	dried, 60°C	578	674	944
modulus	(GPa)		37.4	36.8	79.4
failure strain	(%)		2.4	1.9	1.2
% decrease in properties		65% rel. hum. (20°C, 4 months)	<10	<15	<2
% decrease in properties		boiled (3 weeks)	<60	<15	<4

9.2 MOISTURE

Water uptake

Fibre type and epoxy resin

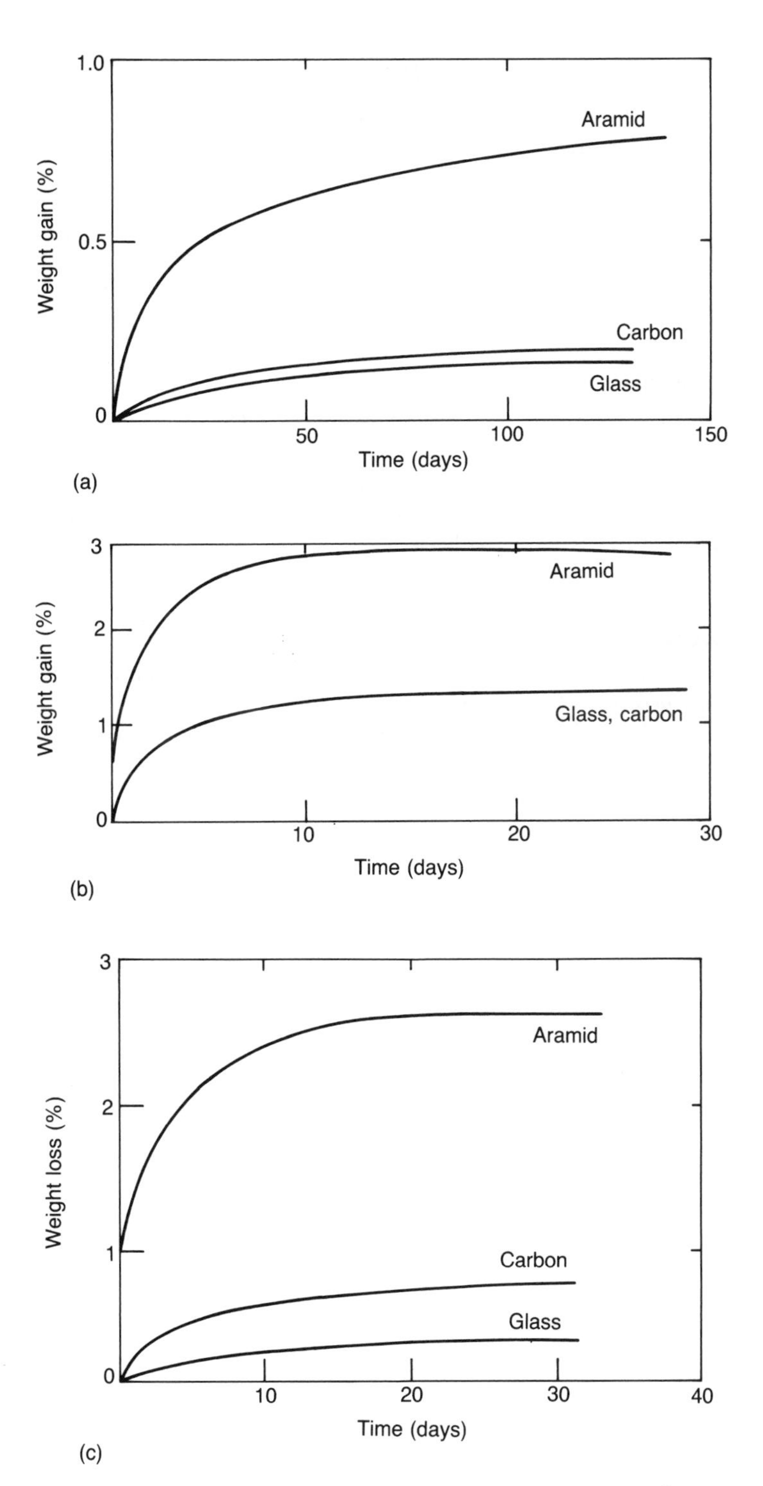

Figure 9.5 Weight change in laminates exposed to various atmospheres; (a) 65% R.H., 20°C; (b) boiling water; (c) drying in air, 60°C (Jones).

■ 9.2 MOISTURE

■ **Water uptake**

Fibre type and epoxy resin

Table 9.3 Effect of preconditioning on interlaminar shear strength of reinforced epoxy laminates

shear strength	treatment	glass	aramid	carbon
initial (MPa)	dried, 60°C	51	22	40
% increase	65% rel. hum., 20°C	14	20	20
% change	boiled	−10	0	+22

Design implications

Humidity does affect composite properties by a small amount; carbon/epoxy is the least affected and aramid/epoxy the most. Since the influence is likely to be exaggerated by sample size (with coupons having a large surface area-to-volume), the effect on large components and structures is likely to be less.

Fatigue results are described in Figure 10.24.

Sources

Jones, C.J. *et al.* (1983) Environmental fatigue of reinforced plastics. *Composites*, **14**, 288–93.

Jones, C.J. *et al.* (1984) Environmental fatigue behaviour of reinforced plastics. *Proc. Royal Society of London*, **396**, 315–38.

9.2 MOISTURE

Shear strength

Aramid, glass, epoxy

Influence on shear strength

The effect of immersion in hot water and subsequent drying has been studied in both aramid/epoxy and glass/epoxy laminates.

Materials

- reinforcement – aramid fibres, Kevlar 49 (du Pont), E-glass fibres (OCF);
- resin – epoxy, Epon 828 (Shell).

Manufacture

Hand lay-up.

Test method

Direct measurement of the interfacial adhesion between a fibre and a resin matrix using the microbond technique.

Observations

Exposure to water at 88°C has a more drastic effect on glass/epoxy than aramid/epoxy. The greater reduction in shear strength (and hence bond strength) of the glass system and its immediate degradation, compared with a one hour 'incubation' period for the aramid system, indicates that the two fibre composites have differing mechanisms of bond deterioration.

Table 9.4 shows that aramid/epoxy also recovers (and indeed improves upon) its shear strength more successfully upon drying following exposure to moisture than glass/epoxy.

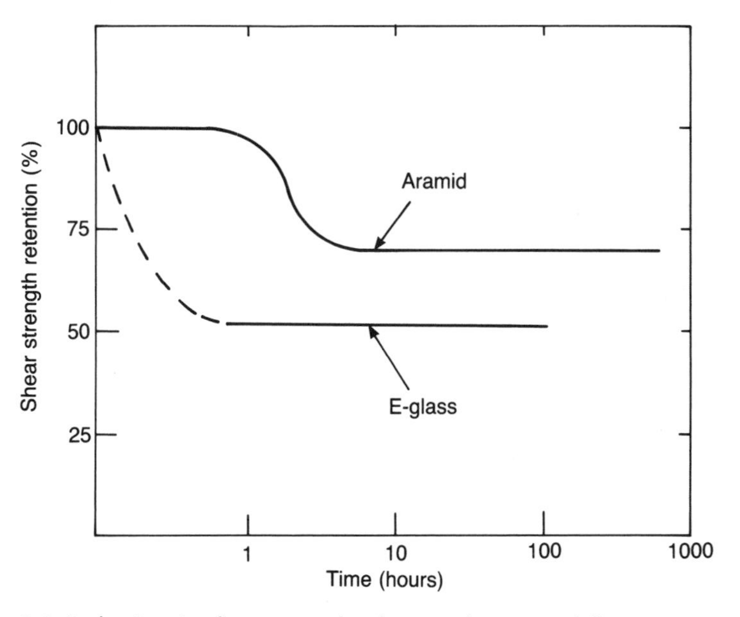

Figure 9.6 Reduction in shear strength of epoxy laminates following immersion in water at 88°C (Gaur and Miller).

■ 9.2 MOISTURE

■ Shear strength

Aramid, glass, epoxy

Table 9.4 Effect of hydro-thermal exposure and subsequent drying on shear strength of reinforced epoxy laminates

		E-glass	aramid	surface-modified aramid
unaged (control)	(MPa)	33.8	31.7	47.4
minimum on ageing in water at 88°C	(MPa)	16.8	23.4	39.7
subsequent drying at 115°C	(MPa)	25.5	34.1	48.5

Surface-modified aramid fibres are the least affected by the ageing/drying process, highlighting the dependence of shear strength on the interfacial matrix bonds.

Design implications

Prolonged immersion can affect properties and shear is a good indication of its influence. As noted in the previous section, coupon data can overestimate the effect on components. It is recommended that suitable size components be tested to determine the influence of size on property degradation.

Source

Gaur, U. and Miller, B. (1990) Effects of environmental exposure on fibre/epoxy interfacial shear strength. *Polymer Composites*, **11**, 217–22.

Effect of stress and glass type

The effect of sustained stress has been measured in air and water for both E glass and E-CR glass laminates.

Materials

- reinforcement – E-glass (OCF), E-CR-glass (OCF);
- lay-up – mat (450)/woven roving (550)/mat (450);
- resin – polyester bisphenol A.

Manufacture

Contact moulding by hand lay-up.

Observations

In both water and air, Figure 9.7, the tensile strength is reduced monotonically under sustained stress for up to 3000 hours. E-CR and E-glass behave identically in air; however, in water E-CR performs very slightly better, initially, but has a lower tensile strength than E glass in the long term.

Design implications

These data provide a limit for the safe working strength of glass/polyester exposed to prolonged stress at 30°C, in either air or water in the principal fibre direction. Other material combinations would need to be tested to check whether they conform to this trend.

If there were significant stresses in other directions then such measurements would need to be repeated with the appropriate coupons. An alternative design approach would be to place sufficient fibres in all principal stress directions to ensure a satisfactory value of creep to ultimate stress.

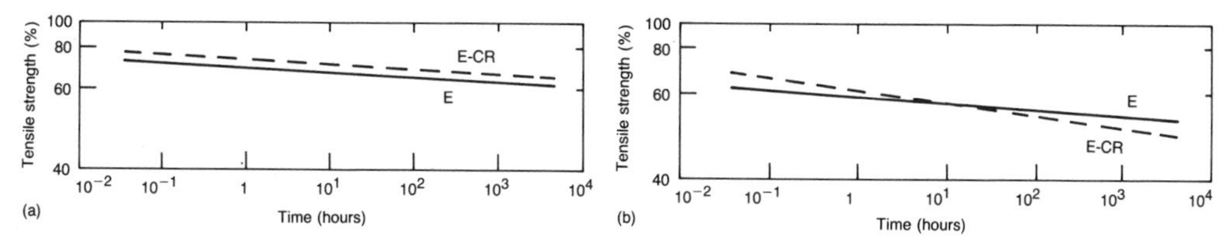

Figure 9.7 Tensile creep of glass laminates at 30°C; (a) in air; (b) in water (Munscheck/OCF).

Source

Munscheck, H. (1987) Prufbericht nr B-8404, IKV, Aachen.
E-CR Glass Update 1 (1991) Owens Corning Fiberglas, Battice.

■ 9.2 MOISTURE

■ **Mechanical properties**

Fibre and resin type

Summary: Effect of Moisture

A useful summary has been devised by Bulder and Bach following a detailed survey of the literature (Table 9.5).

Observations

Glass/epoxy absorbs less moisture than glass/polyester; consequently the property changes are greater with polyester as a matrix than epoxy. Carbon is less affected by water than glass, while glass/carbon hybrids lie somewhere between glass and carbon.

Design implications

Since the resin, rather than the fibre, absorbs the water, the higher the fibre volume fraction the lower water content.

Moisture has an effect, but its influence may only be significant if components are either highly stressed compared with their ultimate strength or used at temperatures close to the heat distortion temperature of the resin. If either condition is applicable then measurements will be needed to determine the safety factor of the design. BS 5480 recommends an operating temperature of at least 20°C below the HDT of the resin for components subject to prolonged exposure to stress.

Table 9.5 Effects of moisture on fibre-reinforced plastics

matrix	max. moist. absorption (weight %)	stiffness change (%)	UTS change (%)	fatigue strength reduction at 10^3 cycles (%)	fatigue strength reduction at 10^7 cycles (%)
glass/polyester	4	−10	−15	−35	−10
glass/epoxy	2	−10	−10	−20	0
carbon/ polyester			−5		
carbon/epoxy	1.5	+1	−2	0	0
glass-carbon/ epoxy	<2	0	−3		

Source

Bulder, B.H. and Bach, P.W. (1991) *Literature Survey on the Effects of Moisture on the Mechanical Properties of Glass and Carbon Plastic Laminates, ECN-C-91-033*, ECN, Petten.

9.3 CORROSION RESISTANCE

The corrosion resistance of reinforced plastics is very dependent upon the type of resin used. As with fire resistance, only indicative information is described herein and more detailed information must be sought from databases and suppliers.

Effect of various chemicals

The chemical resistance of various polymers and glass-reinforced thermosets is shown in Figure 9.8.

Materials

The materials in Figure 9.8 are:

(1) glass-reinforced polyester;
(2) glass-reinforced epoxy;
(3) furane resin;
(4) PVC (rigid);
(5) PE;
(6) PPO.

Observations

Glass/polyester has good resistance to most chemicals except strong bases and strong oxidants. The same is true at 93°C but with slightly decreased resistance. Glass/epoxy shows extremely good resistance to all chemicals except strong oxidants at low temperatures. This resistance decreases as the temperature is increased.

Design implications

Strong oxidants and bases attack the glass/polyester and glass/epoxy most severely whilst the action of organic solvents and acids is resisted with varying degrees of success. Thermoplastics are more resistant against strong oxidants and bases and so could be used as a liner inside a glass/thermoset resin pipe or tank.

Source

Quinn, J.A. (1986) *Design Data Fibreglass Composites*, Owens Corning Fiberglas, Battice.

■ 9.3 CORROSION

■ Chemical resistance

Polymers and composites

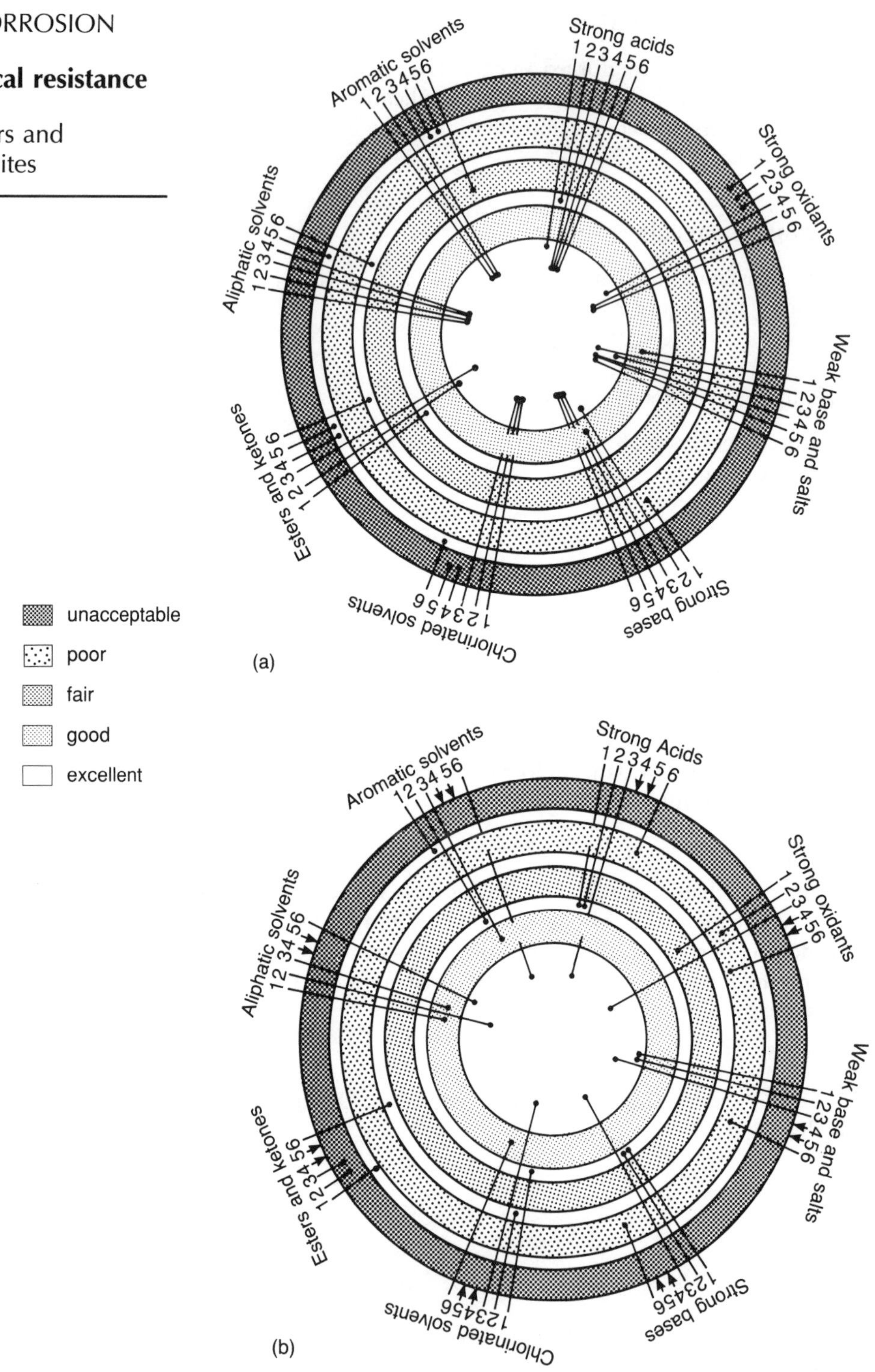

Figure 9.8 Resistance to attack by various chemicals at (a) 23°; (b) 93°C (OCF); materials are identified in the text. Above recommended operating temperature is indicated by an arrow ↑.

Chemical resistance of polyester resins

The chemical resistance of a polyester resin depends upon the constituents from which the resin is manufactured. A general classification is given in Table 9.6.

Alkalis
Not resistant with the exception of iso-NPG, bisphenol and vinyl ester.

Acids
Resistant except at high concentrations in some instances or where the acid is a strong oxidant like H_2SO_4 and HNO_3 above certain concentrations and at elevated temperatures.

Salt solutions
Very resistant and resins can generally be used to within 20°C of their HDT.

Solvents
Although bisphenol, iso-NPG and vinyl ester resins have good all-round chemical resistance, they are not very resistant to aggressive solvents; often highly unsaturated polyester resins are more effective.

Chemical mixtures
Sometimes chemical mixtures, especially solvent mixtures, are more aggressive than the individual solvents themselves; so advice should be sought.

Organic acids
Can behave both as solvents and acids, so extreme care is needed in selection of suitable resins.

Table 9.6 Prediction of the chemical resistance of polyester resins

increasing chemical resistance	influential component in the polyester	classification
↓	phthalic anhydride plus	ortho-polyester
	• diethylene glycol	
	• ethylene glycol	
↓	• propylene glycol	
	iso-phthalic acid plus	iso-polyesters
	• diethylene glycol	
↓	• ethylene glycol	
	• propylene glycol	
↓	• neopentyl glycol	iso-NPG polyester
	bisphenol A	bisphenol-polyester
↓	vinyl ester	vinyl ester

■ 9.3 CORROSION

■ Chemical resistance

Polyester resins

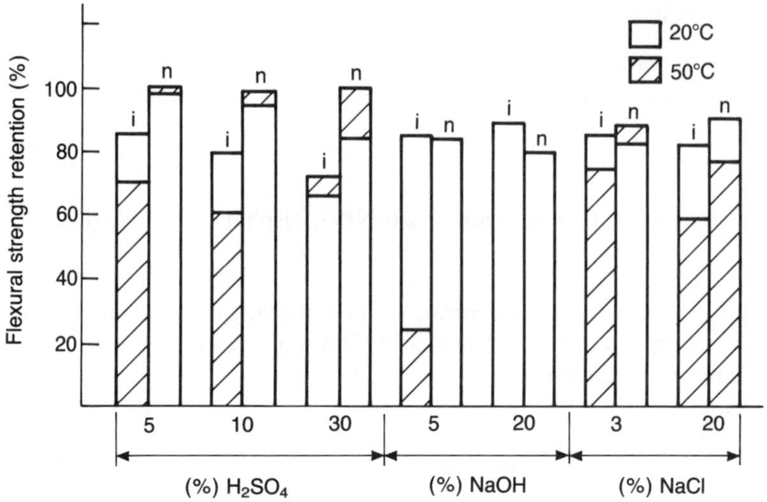

Figure 9.9 Typical cast resin property retention after three months total immersion in chemical environments; i – medium HDT iso-polyester resin; n – high HDT NPG-iso-polyester (all resins post-cured for three hours at 80°C) (Norwood).

Table 9.7 Variation of HDT with post cure temperature for a bisphenol-polyester with an expected HDT of 120°C

cure time (hours)	cure temperature (°C)	HDT (°C)
16	40	60
16	60	85
16	80	107
3	100	114

Some typical data for iso- and iso-NPG polyester resins are given in Figure 9.9 for an alkaline, acidic and saline environment.

To obtain the chemical resistance described above, it is necessary to fully cure the resin. Table 9.7 shows how the HDT of the resin is increased with level of post curing to ensure full cross-linking.

Source

L. Norwood at Scott-Bader.

Effect of acid: fibre type

The effect of various acids on two types of E-glass is given in Figure 9.10.

Materials

- reinforcement – E glass, E-CR glass (OCF);
- resin – beeswax impregnated rovings.

Manufacture

Contact moulding by hand lay-up.

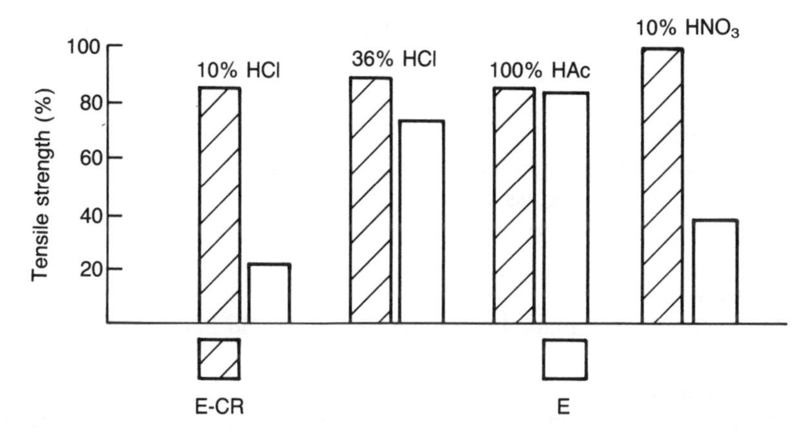

Figure 9.10 Tensile strength retention of beeswax-impregnated rovings after 500 hours of immersion in various acids at 23°C (Renaud/OCF).

The effect of strain corrosion of glass/polyester laminates in dilute sulphuric acid is shown in Figure 9.11.

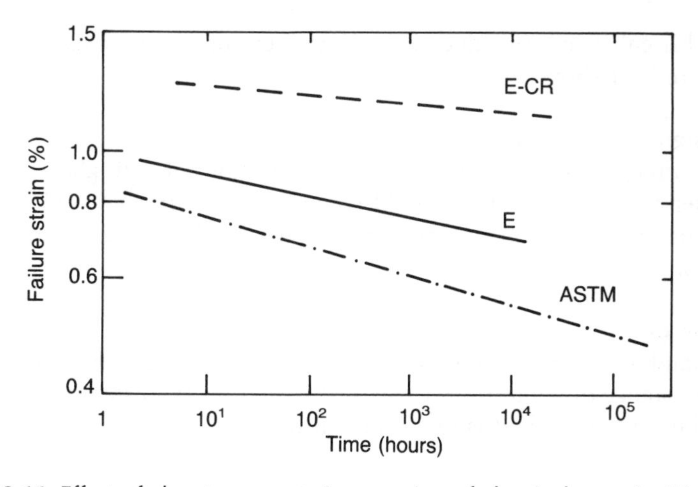

Figure 9.11 Effect of glass type on strain corrosion of glass/polyester in 5% sulphuric acid at 38°C compared with prediction permitted by ASTM D 3681 (Renaud/OCF).

■ 9.3 CORROSION

■ **By acid**

Fibre type

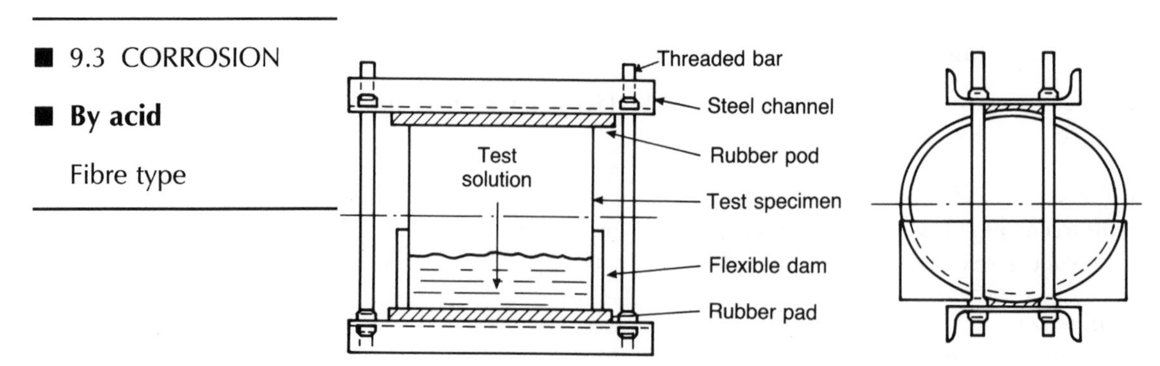

Figure 9.12 Scheme for testing the effect of corrosive liquids on FRP pipe sections under strain induced by compressing a section of pipe using threaded bar (ASTM D 3681).

Materials

- reinforcement – E-glass mat (OCF), E-CR glass mat (OCF);
- surface layer – C-glass veil;
- resin – iso-polyester;

Manufacture

Pipe sections were made by contact moulding and hand lay-up.

Test method

The method for testing pipe sections under strain and in the presence of corrosive liquids is shown in Figure 9.12.

Observations

E-CR glass has better resistance than E-glass to acid attack for a variety of acids when immersed for various periods of time (Figure 9.10). This arises because of the different chemical composition of the two glasses (Table 2.3).

This enhanced resistance is also apparent under stress as the data in Figure 9.11 show.

Design implications

For strength retention under acidic conditions, E-CR glass performs better than E-glass. The ASTM-recommended norm (ASTM D 3681) (Figure 9.11) appears to be conservative for both E-glass and E-CR glass.

Sources

Renaud, C.I. (1986) Study of the influence of glass reinforcement on the resistance of composites to acid corrosion. *Composites*, 3, 101–9.

A similar test method is prescribed in BS 5480.

Effect of acid: resin type

The influence of the resin has been examined by using different types of liner inside a GRP pipe. The maximum strain observed inside the pipe is a good indication of the resistance of the resin to acid attack.

Materials

- reinforcement – (of pipe) E-CR glass (OCF), (of liner) C-glass veil (OCF);
- resin – (in pipe) iso-polyester, (in liner) (a) iso-polyester, (b) vinyl ester, (c) bisphenol A polyester.

Manufacture

Pipes are filament wound.

Observations

The strain-corrosion regression lines have a similar slope for all three liner materials. The strain drops *c*.50% after 20,000 hours. Vinyl ester and bisphenol A liners have better corrosion resistance than iso-polyester in accordance with the classification scheme in Table 9.6.

Design implications

A combination of glass reinforcement and specific resin liners can further improve the long-term strain-corrosion behaviour of pipes in acid environments. In addition, it has been shown that chemical resistant, but brittle matrices have poorer resistance to stress corrosion than tougher flexibilized matrices such as vinyl ester.

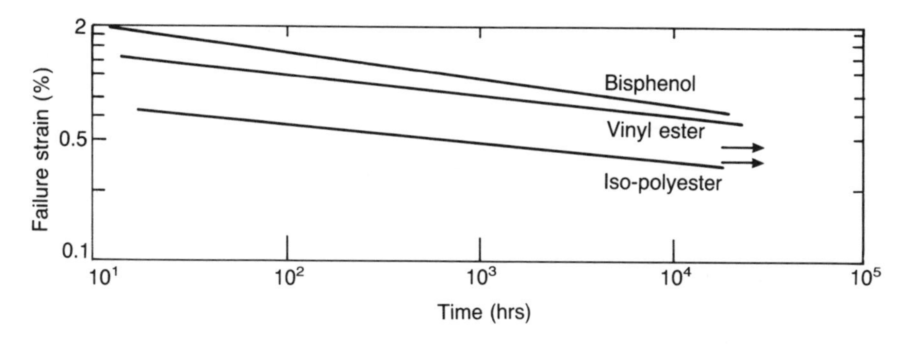

Figure 9.13 Effect of using different resins in the liner of pipe rings on their strain-corrosion in 5% sulphuric acid (Hogg and Hull).

Source

E-CR Glass Update 1 (1991) Owens Corning Fiberglas, Battice.

Hogg, P. and Hull, D. (1982) *Proc. British Plastics Federation Conference.*

■ 9.3 CORROSION

■ **By alkali**

Glass/polyester

Effect of alkali

The effect of an alkali on the strain corrosion of a pipe is shown in Figure 9.14.

Materials

- reinforcement – E-glass or E-CR glass, chopped strand mat and woven roving (OCF);
- resin – bisphenol A polyester;

Manufacture

Contact moulding and hand lay-up.

Test method

Tensile stress corrosion, component immersed in NaOH at 30°C.

Observations

The retention of strength during tensile stress corrosion in alkali decreases linearly with time on a log–log axis (Figure 9.14). The regression lines giving the failure times under sustained stress show no significant difference between the two types of E-glass.

Design implications

Properties of glass/polyester are appreciably worse in the presence of alkalis than acids (*cf.* Figures 9.14 and 9.11 for example). Epoxy and furane would be better matrices (Figure 9.8) whilst the use of a thermoplastic liner could also be considered.

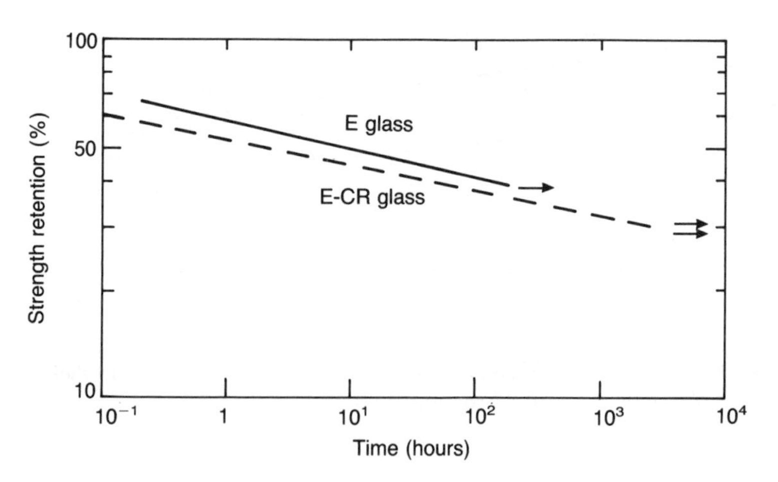

Figure 9.14 Tensile stress-corrosion of laminates in sodium hydroxide at 30°C (Munscheck/OCF).

Source

Munscheck, H. (1987) Prufbericht nr B-8703, IKV, Aachen.

9.4 HARDNESS, ABRASION AND WEATHERING

Three types of hardness are of concern:

- ability to resist deformation under the influence of an indentor;
- ability to resist scratching;
- ability to scratch other materials.

Barcol hardness

Barcol hardness is the standard method of measuring hardness in resins (both unreinforced and reinforced). It is primarily used to determine the state of cure of a resin, but it can also be used to assess the strength of a resin and possibly a composite (Figure 9.15).

As discussed in Chapter 12, care must be taken with this technique in making the measurements and analysing the results.

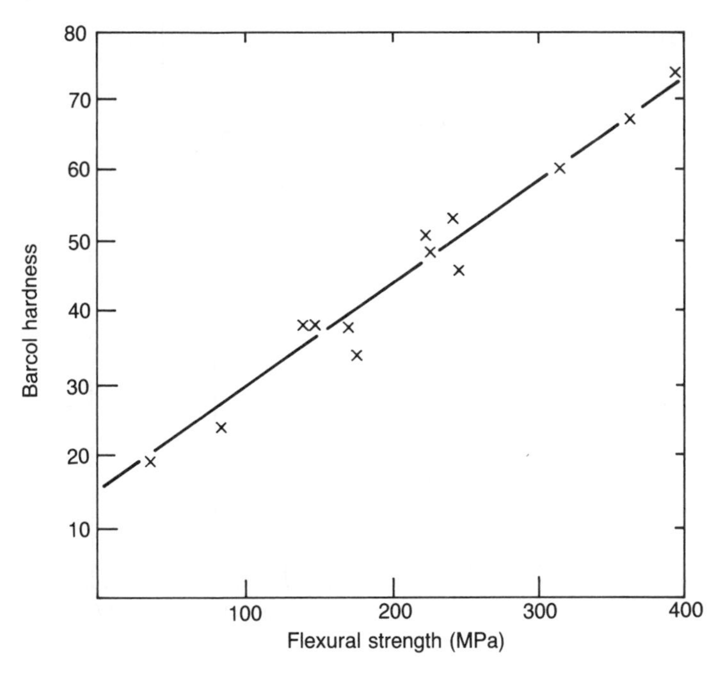

Figure 9.15 Relationship between average Barcol hardness and flexural strength for a variety of balanced glass fabric-reinforced polyester laminates (Mayer).

Source

Mayer, R.M. (1985) Material selection and quality control of glass reinforced plastics, *Proc. 7th British Wind Energy Workshop, Oxford*, MEP, London.

■ 9.4 ABRASION

■ **Scratch**

Scratch resistance

Two types of test are in use to compare various types of material,

- a hardness test in which a loaded indentor is drawn across the surface (BS 3400 part E2);
- an adhesion test for checking the integrity of paint layers, surface coatings and gel coats to an underlying surface (BSAU 148 part 3).

Abrasion resistance and wear

Abrasion resistance tests are generally undertaken using a rotating wheel lapping a surface. Some comparative data for resins are given in Figure 9.16.

It is clear that there is a wide spread in resistance with polyethylene being the best and epoxy the least resistant.

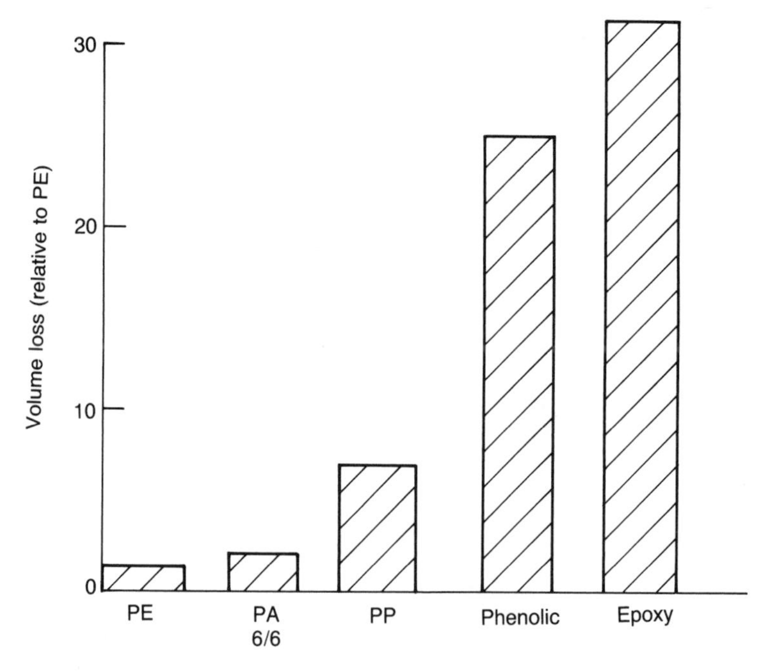

Figure 9.16 Comparative abrasion resistance of various resins; polyethylene is of ultra-high molecular weight type (Stein).

Source

Stein, H.L. (1987) Ultra-high molecular weight polyethylene, in *Engineered Materials Handbook, Volume 2: Engineering Plastics*, ASM International, Metals Park, Ohio.

Effect of fibre fraction

Glass, being a hard material, will alter the abrasion resistance of the composite. The effect of fibre fraction is shown in Figure 9.17.

Materials

- reinforcement – E-glass (Silenka);
- resin – epoxy, LY554/HY956 (Ciba-Geigy).

Manufacture

Pultruded rods, 10 mm diameter.

Test conditions

Dry wear test by abrading a pin on a disc.

Observations

Increasing fibre volume fraction results in greatly increased wear resistance up to 30 v/o. Above 30 v/o, the decrease in wear resistance is attributed to insufficient resin to bond the fibres together. Additional tests demonstrate that increasing the sliding velocity increases wear resistance, and increasing contact pressure decreases it.

Design implications

Glass fibres are beneficial in reducing wear rate provided that the fibre fraction is less than 30 v/o. The interfacial bond between matrix and fibres plays an important role in the wear process. To ensure that cracks do not initiate and propagate in the matrix, leading to failure in these

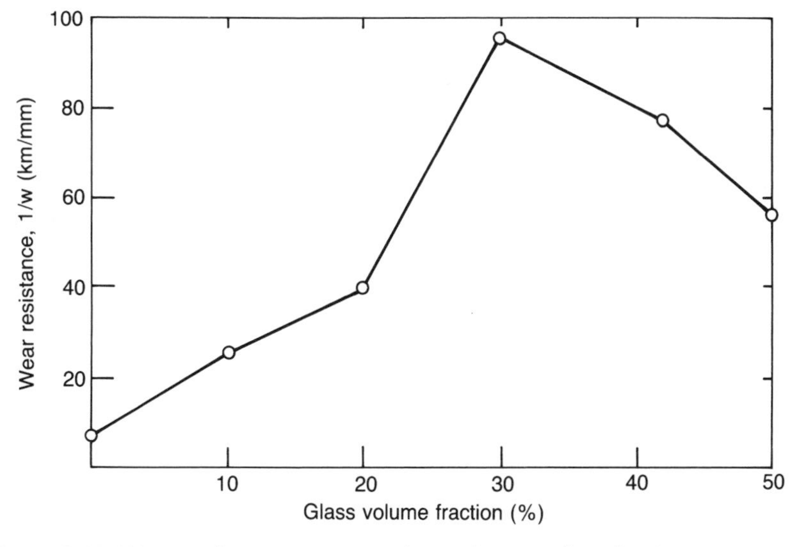

Figure 9.17 Wear resistance versus volume fraction for glass/epoxy composites (Zamzam).

■ 9.4 ABRASION

■ **Fibre fraction**

Glass/epoxy

bonds, the appropriate fibre/resin ratio should be selected to give optimum wear rate.

Source

Zamzam, M.A. (1990) The wear resistance of glass fibre reinforced epoxy composites. *Journal of Materials Science*, **25**, 5279–83.

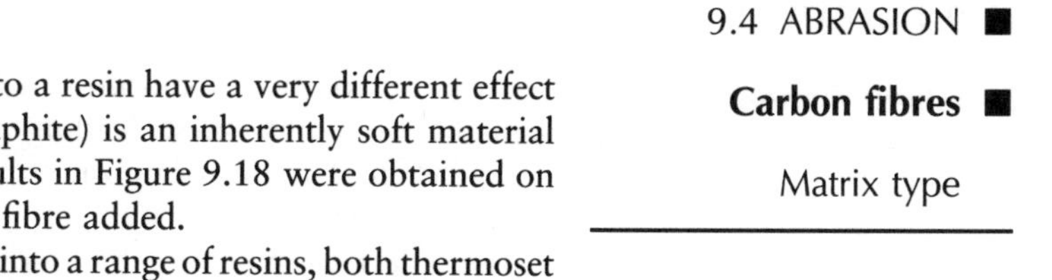

9.4 ABRASION ■

Carbon fibres ■

Matrix type

Influence of carbon fibres

Carbon fibres when incorporated into a resin have a very different effect from glass as carbon (in reality, graphite) is an inherently soft material with lubricating properties. The results in Figure 9.18 were obtained on a pair of samples with and without fibre added.

The incorporation of carbon fibres into a range of resins, both thermoset and thermoplastic, is very beneficial in terms of wear rates. Wear can be reduced by as much as a thousand-fold, thus forming the basis of a class of materials which are self-lubricating.

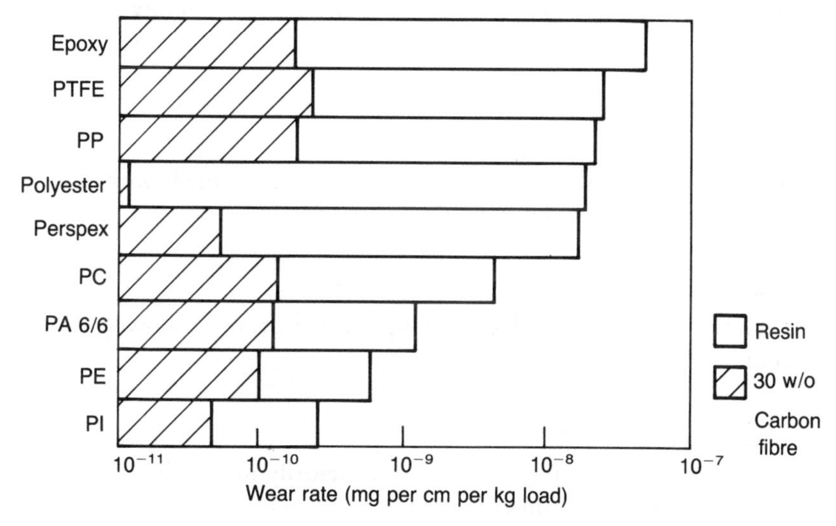

Figure 9.18 Effect of carbon fibres on the wear rate of polymers (Lancaster/RAE).

Source

Lancaster, J.L. (1966) Effect of carbon fibre reinforcement on the friction and wear of polymers, *Report 66378*, RAE, Farnborough.

■ 9.4 WEATHERING

■ **Glass/phenolic**

Weathering

Weathering (outdoor exposure) encompasses the effects of moisture and sunlight on composites, leading to hydrolytic and photo-oxidative degradation. The most important of these is generally the effect of the ultra-violet (UV) rays of sunlight.

The effect of moisture has already been considered. Whereas UV degradation has been a problem in the past, UV-stabilizers have now been developed so that degradation is no longer a prime concern. Fading of dyes used for pigmentation can occur and this would need to be checked if deemed necessary.

As it is not possible to pigment phenolic resins successfully, an external paint layer has to be applied. Tests to check the durability of this layer are listed in Table 9.8.

Materials

- reinforcement – glass CSM;
- resin – phenolic J2027 (BP Chemicals), painted with two-part polyurethane coating (Trimite Paints Ltd).

Manufacture

Contact moulding and hand lay-up, epoxy primer followed by a top coat.

Observations

A polyurethane coating performs satisfactorily under all test conditions provided a suitable primer is used.

Table 9.8 Durability of a polyurethane paint system on a phenolic laminate

test	method	result
2000 h weatherometer	BS 3900 F.3	no colour change, chalking or loss of gloss
250 h 100% humidity	BS 3900 F.2	no blistering or gloss change
500 h warm salt spray	ASTM B 117	no blistering or gloss change
500 h cold salt spray	BS 3900 F.4	no blistering or gloss change
500 h fresh or salt water immersion		no blistering or gloss change
abrasion resistance	BSAU 148 part 4 (Taber Abrader)	CS17 wheel, 1 kg load loss 0.1 g after 1000 revs
adhesion test, cross hatch	BSAU 148 part 3	excellent
scratch hardness	BS 3400 part E2	passes 2000 g

Design implications

A paint system is the only way to achieve a tough, durable and coloured surface for phenolic laminates. Specific applications should be checked with manufacturer.

Source

Forsdyke K.L. and Hemming, J. (1989) Phenolic GRP and its application in mass transit, *Proc. Annual Conference, Composites Institute*, SPI.

■ 9.4 WEATHERING

■ **Sheet mouldings**

Weathering of sheet mouldings

The weathering of a sheet moulding compound (SMC) has been investigated after exposure up to a period of 18 months outdoors in a sun-facing direction.

Materials

SMC comprising 25 w/o glass and ortho-polyester resin.

Manufacture

Press moulding.

Observations

There is an observable decrease in both modulus and strength during the summer months. In Figure 9.19 the stiffnesses show a dramatic drop-off during the summer, particularly flexural moduli as flexural loading is more responsive to the state of the surface.

Design implications

This type of SMC does not exhibit inherent resistance to weathering, having constituents which are susceptible to both oxidative and hydrolytic degradation. Surface protection, such as painting, would be needed if the material was to be outdoors longer than a few months.

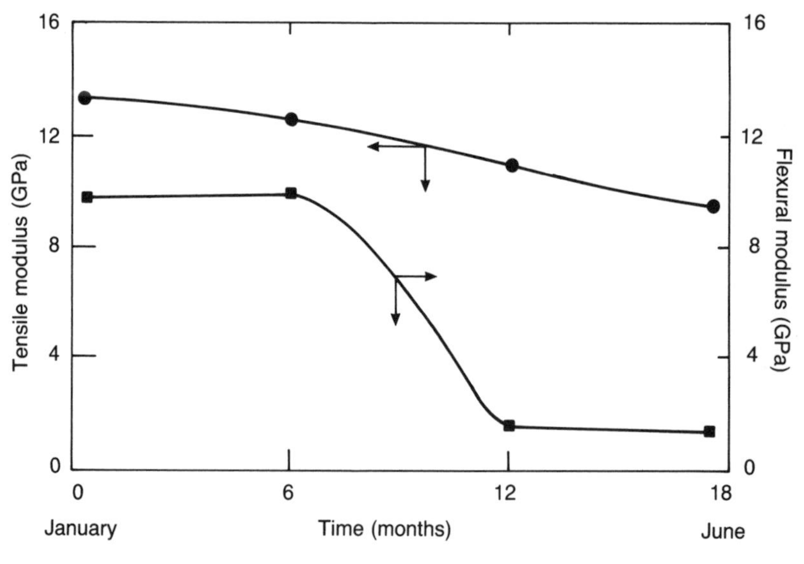

Figure 9.19 Degradation of moduli by outdoor exposure (Roylance and McElory).

Source

Roylance, D. and McElroy, P. (1989) Weathering of polymer composites, *Composites Asia Pacific Conference.*

9.5 DESIGN STRATEGY

The type and quality of the surface is of prime importance in withstanding environmental attack.

The quality of the gel-coat is all important in controlling humidity ingress and preventing blister formation and osmosis. The latter has been of particular concern with boat hulls exposed to sea water. With the aid of suppliers, moulders have developed suitable manufacturing techniques.

Coupon data can be used to assess the effectiveness of various materials and techniques, but extrapolation of such data to large components and structures is uncertain.

To prevent corrosion, both resin and fibre should be selected to be as resistant as possible taking into account other service requirements. It is essential for the resin to be fully cured. In addition, the structural laminate needs to be adequately protected from the environment by a substantial barrier layer.

This could consist of a GRP layer several millimetres thick made up of a synthetic tissue or C-glass veil and chopped strand mat at a high resin/glass ratio (Figure 9.13). Alternatively a suitable liner (generally thermoplastic) should be considered.

Accelerated corrosion tests should be carried out to determine the effectiveness of the barrier or liner.

Abrasion resistance can be achieved by suitable material selection and simple wear tests.

Weathering to a very large extent can be controlled by the use of a suitable gel-coat and the incorporation of UV-stabilizers within the resin.

9.6 REFERENCE INFORMATION

Information is primarily available about corrosion resistance:

Design Data of Fiberglas Composites (1986) Owens Corning Fiberglas, Ascot – a summary table of chemical resistance to polyester laminates and other plastic materials.

Scott-Bader, Woolaston, Northamptonshire, NN9 7RL, UK, Tel. (44) 0933 663100 – provide a good list of the maximum working temperature for exposure to a large range of chemicals for six principal types of glass-reinforced polyester resin.

Chemres II is a materials selection database supported by RAPRA Technology, Shawbury, Shrewsbury, Shropshire SY4 4NR, UK, Tel. (44) 0939 250383 – information provided on the chemical resistance of 44 plastics and 19 rubbers in 192 chemical environments.

E-CR Glass Update 1 (1991) Owens Corning Fiberglas, Battice – a comprehensive review of the existing data and trends for E-glass and E-CR glass exposed to various environments.

10 CREEP AND FATIGUE

SUMMARY

The long-term response of materials to load in either a steady or fluctuating form is considered. The role of the reinforcement in resisting the load is considered as well as the role of the resin in protecting the reinforcement and transferring the load to the fibres.

10.1 INTRODUCTION

Long-term loading can either be steady (creep) or fluctuating (fatigue) or a combination of both. Both creep and fatigue are characterized by three stages of damage accumulation, which could eventually lead to failure (Figure 10.1).

In stage I the component is initially stressed and some damage is introduced as may be detected by a drop in stiffness.

Stage II takes place over a much longer time period in which there is only a slow change in strain and stiffness with time ('steady state').

Stage III is characterized by an ever-increasing amount of damage, which could lead to failure. This can usually be detected by an increasing drop in stiffness. If the stress is very low then this stage may only be reached after a very long time if at all (fatigue or creep limit).

What constitutes failure (section 1.4) will be dictated by the design, but a 10% drop in stiffness is becoming accepted in fatigue. For creep, a strain limit is usually imposed by the design.

Material selection

In general, the fibre is chosen for its ability to withstand load and the resin for its temperature capability and resistance to environmental and corrosive attack.

The resin is generally the weaker constituent of a composite, and so will set the upper limit to the service temperature. This arises because of its inability to transfer the load effectively as the service temperature approaches the heat distortion temperature of the resin. As discussed subsequently, it is for this reason (amongst others) that the acceleration of creep measurements by undertaking shorter tests at higher temperatures (time–temperature superposition) must be undertaken with caution.

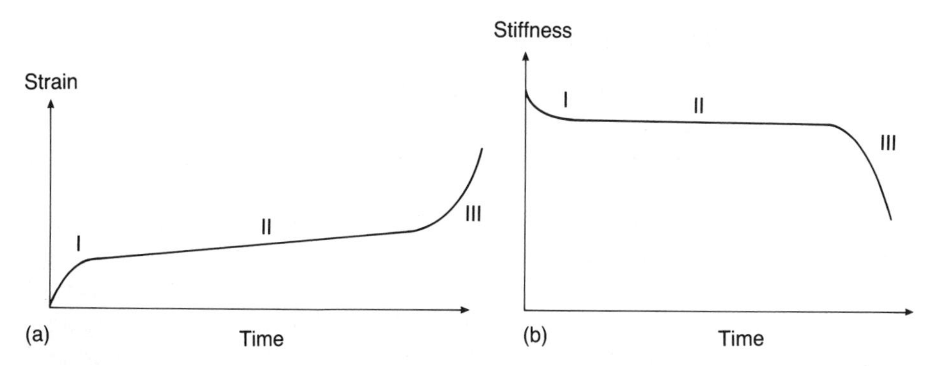

Figure 10.1 (a) Typical variation of creep strain with time; (b) typical decrease in stiffness with time for fatigue loading.

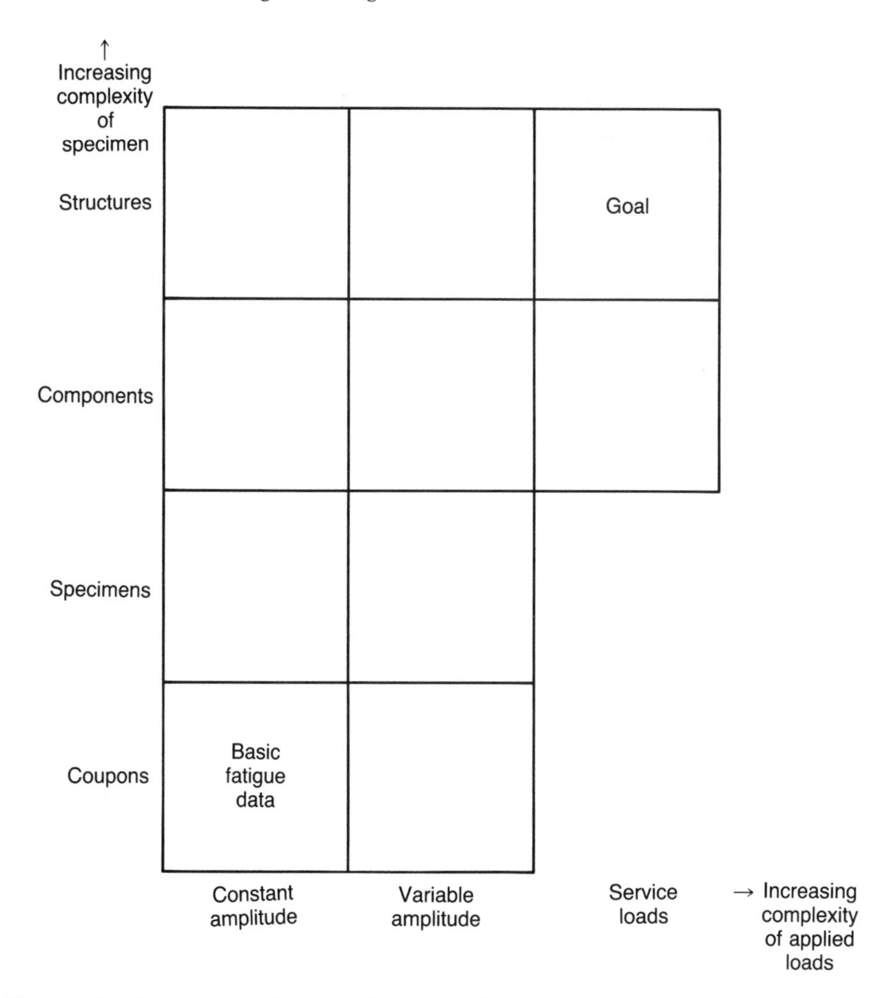

Figure 10.2 Design considerations for structural use under fatigue loading (Haibach).

Complex loading and component size

The incorporation of composite materials into load-bearing components and structures requires a perception of both load complexity and the structural response to load. This is illustrated schematically in Figure 10.2.

The designer has to commence with the basic data collected on coupon specimens and determination of the appropriate service loads of the component. If a high level of integrity is required then the effect of variable loading rather than constant loading will have to be investigated.

Having completed the design, the component can be manufactured and tested under the appropriate service loads. Noteworthy examples of this approach include glider wings, wind turbine blades and leaf springs.

Source

Haibach, E. (1981) Fatigue data for design applications, in *Materials, Experimentation and Design in Fatigue*, *Proc. Fatigue '81*.

10.2 CREEP

A number of creep studies have already been discussed in Chapter 9 on environmental resistance. These include composites involving E-glass or E-CR glass in air, water and chemical environments (Table 10.1).

The most significant aspect of this data is that E-glass/polyester can withstand 50% of its ultimate stress for 100,000 hours at 30°C (Figure 9.7). Since composites tend to be stiffness limited rather than strength limited, this magnitude of creep loading is unlikely to arise. The creep strength is diminished by environmental attack, exposure to alkali being the most severe (Figure 9.14).

Table 10.1 Effect of environment on creep

glass type	resin	medium	temperature (°C)	loading	reference (Figure)
E E-CR	polyester	air water	30	tensile	9.7
E E-CR	polyester	5% H_2SO_4	38	tensile	9.11
E-CR	polyester vinyl ester bisphenol-polyester	5% H_2SO_4		compressive	9.13
E E-CR	bisphenol polyester	10% NaOH	30	tensile	9.14

■ 10.2 CREEP

■ **Fibre type**

Vinyl ester resin

Effect of fibre type

The creep rate of the principal fibre types has been compared by testing in tension at 50% of their breaking strength (Figure 10.3).

Materials

- reinforcement – E-glass (Pilkingtons), aramid, Kevlar 49 (du Pont), carbon XAS (Hysol Grafil);
- resin – vinyl ester, Derakane 470-36 (Dow Chemicals);
- steel – low relaxation (roping type) and normal.

Manufacture

By pultrusion, rods 5 mm diameter (composites), similar diameter steel rods.

Observations

This data set shows that aramid rod has the highest creep rate with E-glass rod lower and carbon rod the lowest. Standard roping steel is comparable with E-glass and low relaxation steel with carbon. The properties are compared in Table 10.2.

The strength of aramid and carbon rods compares favourably with that of roping steel on an absolute basis, but their mass per unit length is very much less.

The creep of aramid (Kevlar 49) and S-glass has also been studied in the form of resin-impregnated strands (Figure 10.4).

The data show that the creep rate of Kevlar 49 is less than that of S-glass, whose creep behaviour should be similar to E-glass. This differs from the data in Figure 10.3, but whether this arises because of the nature of the test would require further investigation.

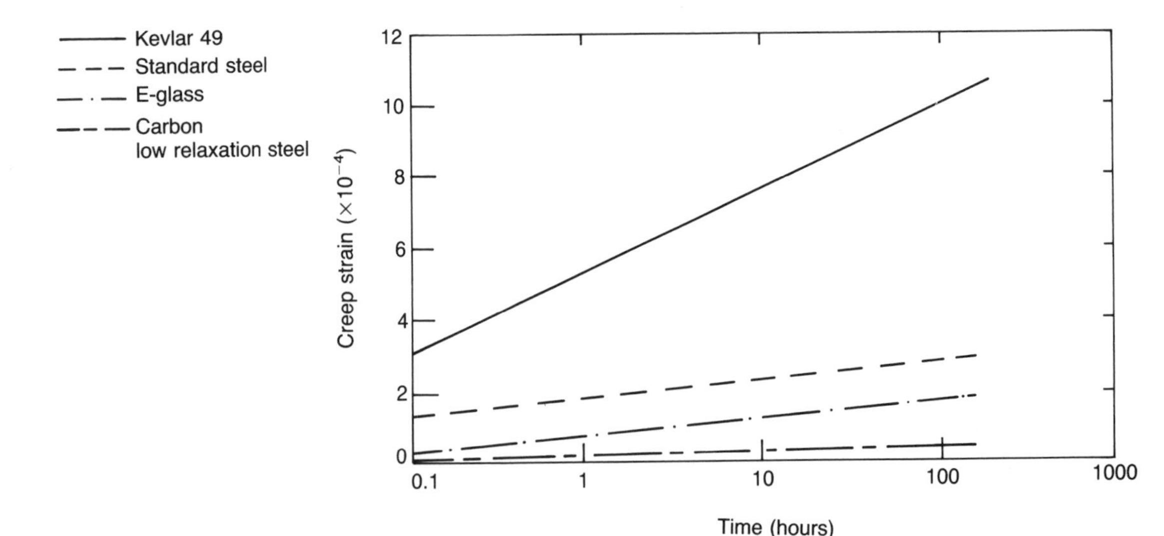

Figure 10.3 Tensile creep properties of pultruded rods and steel rods tested at 50% breaking strain and 20°C (Yeung and Parker).

10.2 CREEP ■

Fibre type ■

Vinyl ester resin

Table 10.2 Properties of rods

property		E-glass	aramid	carbon	roping steel
fibre fraction	v/o	61	63	63	
density	Mg/m^3	2.02	1.36	1.57	7.86
tensile strength	MPa	1060	1590	1610	1820
tensile modulus	GPa	45.7	64.3	136	200
elongation at break	%	2.3	2.5	1.2	3.2
mass/unit length	g/m	40	27	31	154

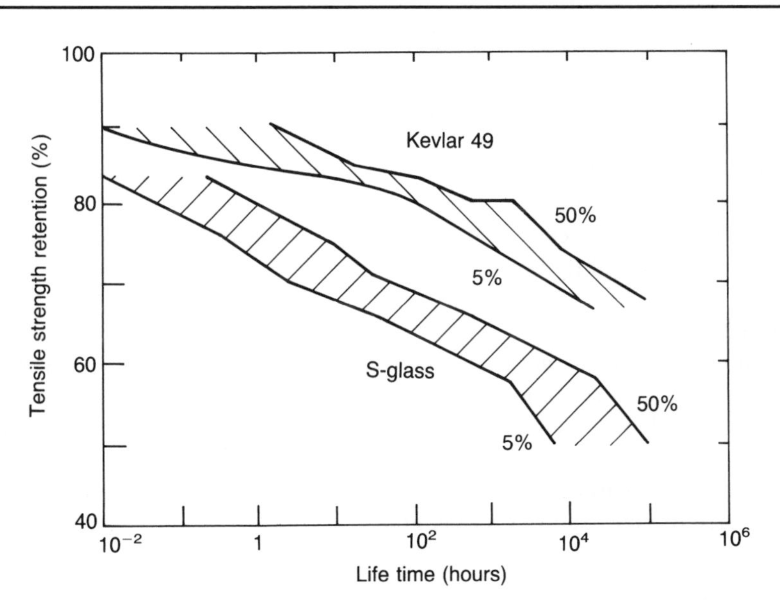

Figure 10.4 Creep response of aramid (Kevlar 49) and S-glass at room temperature (du Pont).

Design implications

Creep is low in fibre-reinforced plastics at room temperature and comparable with or better than steel depending upon the type of fibre and steel. The stiffest fibre (carbon) has the lowest creep rate.

Subsequent fatigue studies on twisted strands show that carbon and aramid composite strands had better fatigue resistance than roping steel.

It is for these reasons that such materials are of interest for tension members in applications like mooring ropes and very long span bridges.

Sources

Yeung, Y.C. and Parker, B.E. (1987) Composite tension members for structural applications. *Composite Structures*, **4**, 1309–19.

du Pont trade data.

■ 10.2 CREEP

■ **Resin type**

Chopped strand mat

Effect of matrix

A comparison of glass/polyester and glass/phenolic when exposed to water is given in Figure 10.5.

Materials

- reinforcement – chopped strand mat, 450 g/m^2 (Pilkingtons);
- resin – phenolic, Cellobond J2018L (BP Chemicals).

Manufacture

Contact moulding, $V_f = 21\%$.

Test conditions

Immersion in deionized water, temperature 25°C.

Observations

The lines in Figure 10.5 are drawn as the best fit through the data which exhibit considerable scatter particularly for the polyester matrix at short times. The properties appear comparable at short times, whilst at failure times in excess of 10^6 s the working stress of the phenolic composites is slightly lower.

Design implications

Phenolic appears to be a suitable alternative as a matrix to polyester resin under such conditions. However, there are fewer measurements on the former than the latter so further tests would be required to validate such results.

Since water can attack the fibre/resin interface, selection of a suitable fibre size to give a good bond is important (Figure 4.11).

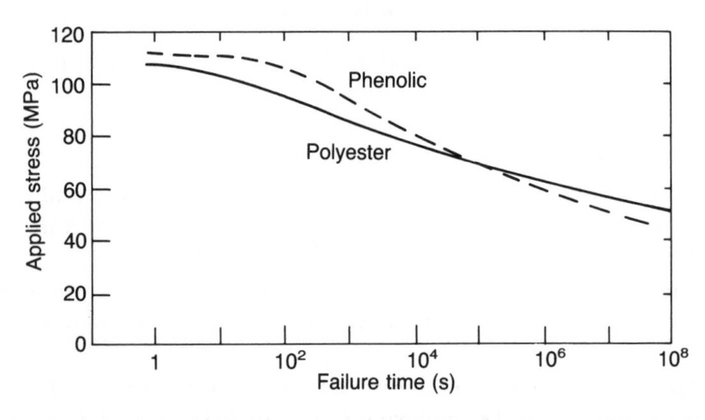

Figure 10.5 Creep-rupture of CSM laminates in deionized water at 25°C (Aveston/NPL).

Source

Aveston, J. *et al.* (1985) *Long-term Strength of Glass Fibre/Phenolic Resin Composites, Report DM(A) 98*, National Physical Laboratory, Teddington.

Effect of temperature

Materials
- reinforcement – unidirectional glass fabric Equerove 23/47 (Pilkingtons);
- resin – iso-polyester, A283/270 (BP Chemicals).

Manufacture
Contact moulding.

Test conditions
Liquid immersion in water, temperatures of 30°C, 45°C and 60°C.

Observations
Failure at 45°C took longer than at 30°C but at 60°C took least time. This is because increasing the temperature increases both the diffusion coefficient of the liquid in the matrix and also the fracture toughness of the resin. The effects of these parameters are opposing.

Design implications
There is no evidence of any stress limit in these data. Because diffusion of the liquid to the interface between fibre and resin is important, measurements would have to be made with other types of liquids. Moreover, such a behaviour might allow a time–temperature superposition to be applied (Figure 10.10).

The kinetics of water uptake and their effect on tensile and shear properties were discussed in Chapter 9, section 9.2.

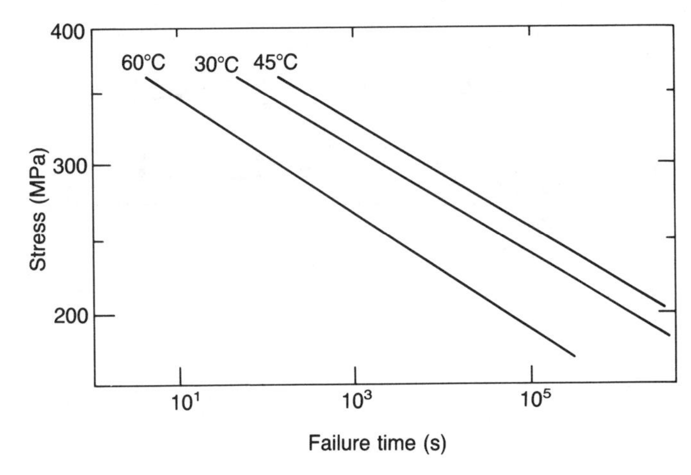

Figure 10.6 Linear regression lines for creep-rupture times at 30°C, 45°C and 60°C (Pritchard and Speake).

Source
Pritchard, G. and Speake, S.D. (1988) Effects of temperature on stress-rupture times in glass/polyester laminates. *Composites*, **19**, 29–35.

■ 10.2 CREEP

■ **Strain**

Glass polyester

Effect of strain

Materials

- reinforcement – combination fabric, balanced woven fabric plus CSM, $V_f = 30\%$;
- resin – polyester Crystic 625 (Scott-Bader).

Manufacture

Contact moulding.

Test conditions

Temperature 60°C, liquid immersion in water.

Observations

Due to the direct relationship between initial strain and applied load, a linear increase in time to failure was expected as initial strain decreased. However, time to failure increases without limit as the initial strain approaches 0.8% – thus setting a limit below which no further creep was detected.

Design implications

If the creep strains can be kept below the creep limit through good design then this would be the ideal design approach. It is essential to check through whether this condition will occur for a specific material combination and loading.

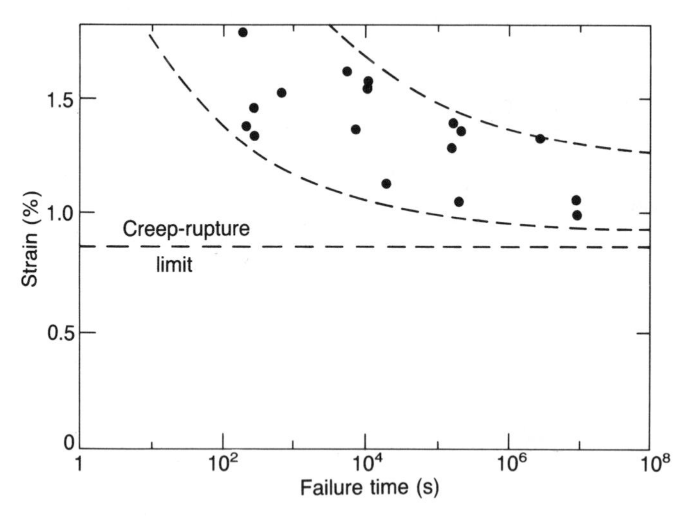

Figure 10.7 Relationship between initial tensile strain and time to failure for glass polyester composite (White and Phillips).

Source

White, R.J. and Phillips, M.G. (1985) Environmental stress rupture mechanisms in glass fibre polyester laminates, in *Proc. International Conference on Composite Materials*, 1089–99.

Effect of transient overloads

Materials

- resin – polyester, 625 TV (Scott-Bader);
- fabric – glass, CSM(300)/W(830)/CSM(300) (woven rovings in hoop and axial directions of pipe) $V_f = 30\%$.

Manufacture

Contact moulding of pipe sections.

Test conditions

Water immersion at 40°C at uniform internal pressure.

Observations

Pipes were pre-treated to transient overloads (*c*.60% of UTS) and then creep tested along with virgin pipes. The pipes that had been exposed to overloads failed at shorter times than the non-overloaded pipes.

Design implications

Transient overloads create damage which can cause premature failure; they thus need to be considered during the design stage. If these cannot be eliminated by design then the tests should be carried out to determine the reduction in life. Examples of such overloads include water hammer in cooling pipes (as above) and gust turbulence for wind turbine blades.

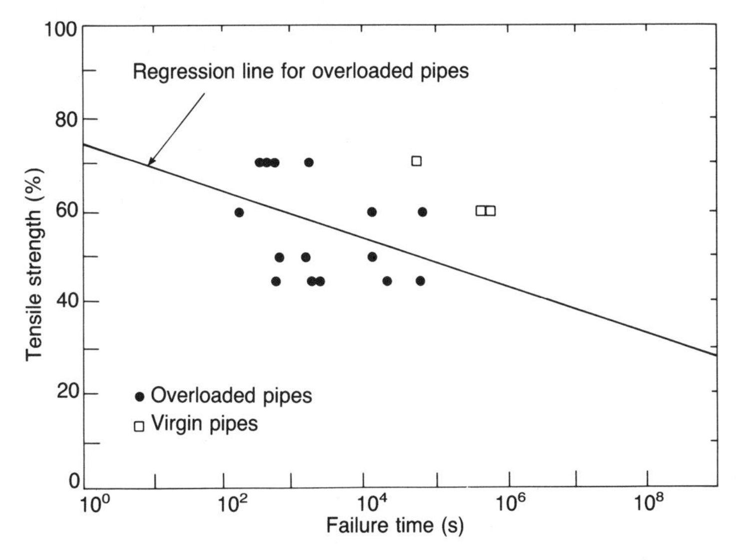

Figure 10.8 Stress-rupture behaviour of GRP pipe sections (Rawles/ASME).

Source

Rawles, J.D. *et al.* (1990) The effect of transient overloads on the long term performance of GRP pipes from power station cooling water systems, *Proc. ASME Pipes and Pressure Vessels Conference*, ASME, Nashville, Tenn.

■ 10.3 TIME–TEMPERATURE

■ **Creep rates**

Glass / vinyl / ester

10.3 TIME–TEMPERATURE SUPERPOSITION

The correlation between time and temperature is illustrated in Figures 10.9 and 10.10.

Materials

- reinforcement – glass rovings;
- resin – vinyl ester, Derakane 470-36 (Dow Chemicals).

Manufacture

Filament wound plates, 45° to axis.

Observations

The creep rate under shear loading increases with both time and temperature. Similar curves are obtained for tensile creep and other fibre angles. Using a transformation equation, it was possible to superpose these curves to form a master curve.

Design implications

This method consists of taking creep data at high temperatures and short times and constructing a master curve. This could then be used to allow extrapolation to lower temperatures and longer times and hence give an indication of the creep response. It may only apply for certain load conditions and specific material combinations and needs to be validated.

If creep is a design constraint then suitable component tests will need to be undertaken. BS 5480 recommends that materials should only be

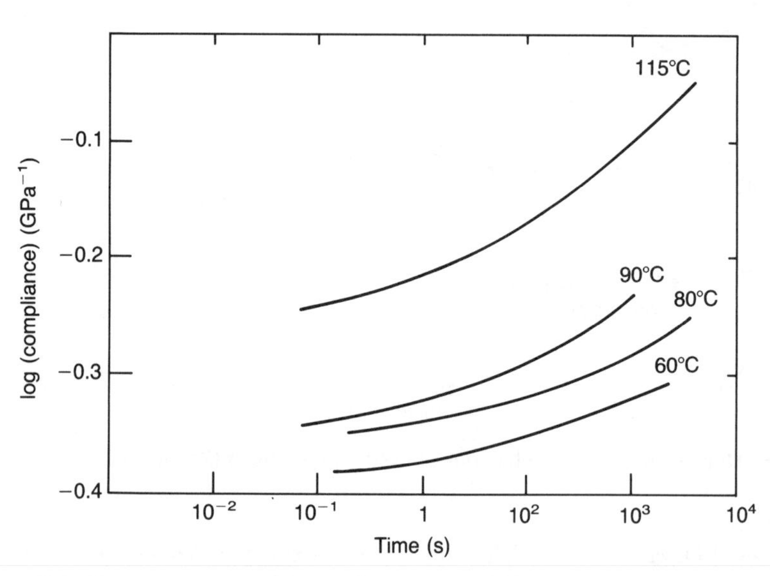

Figure 10.9 Shear creep compliance of ±45° glass laminates at various temperature (Sullivan).

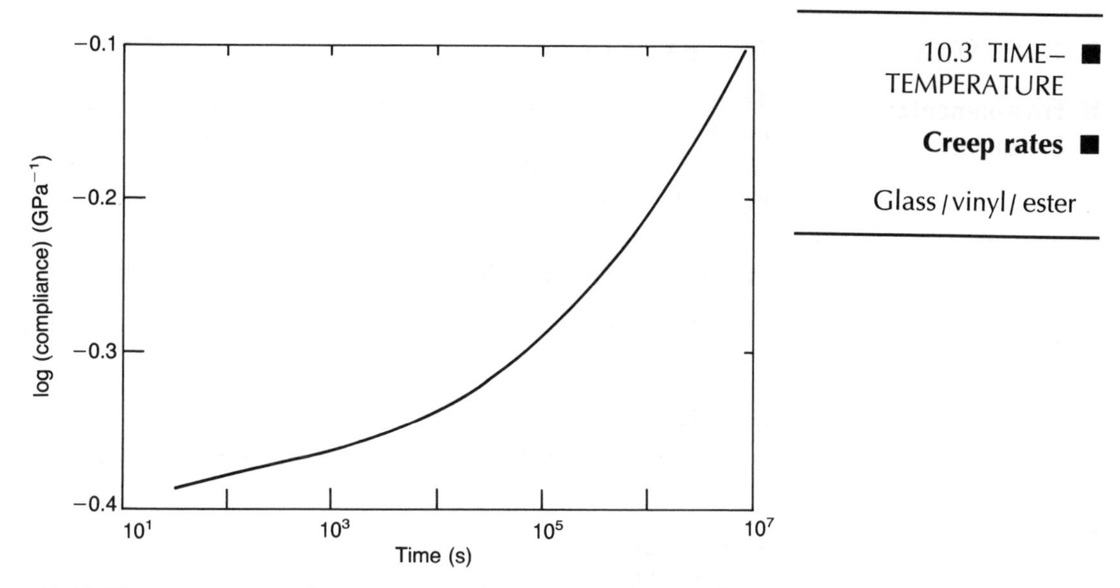

Figure 10.10 Time–temperature shear creep compliance master curve (Sullivan).

used in service at temperatures of 20°C or less below their heat distortion temperature.

Source

Sullivan, J.L. (1990) Creep and physical aging of composites. *Composites Science and Technology*, **39**.

■ 10.4 FATIGUE

■ **Environmental factors**

Glass/ epoxy

10.4 FATIGUE

There are many aspects to consider, which include type of loading, load duration, load introduction, loading rate, temperature and environment. The major influences are considered in turn with a suitable generic illustration.

Influence of fibre fraction

The influence of fibre fraction is illustrated in Figure 10.11 for a woven fabric.

Materials

- reinforcement – plain weave glass fabric, mass 200 g/m^2;
- resin – epoxy.

Manufacture

Press moulding.

Test conditions

Load control, test frequency = 8.3 Hz, minimum/maximum stress (R = 0.11), tensile loading.

Observations

The fatigue data cover the fibre fraction from 20 v/o to 50 v/o and lines of constant life are plotted. Though the static strength increases linearly with fibre fraction up to 50 v/o, the fatigue strength goes through a maximum *c*.40 v/o and then decreases.

Design implications

The rule of mixtures (section 1.3) only applies below the optimum fibre fraction for static properties as can be observed in Figures 4.8 and 5.6. An optimum also exists for fatigue properties and this lies around 40 v/o for this fabric.

The existence of an optimum fraction needs to be established if a high level of properties is desired. Some fabrics pack better than others. If these cannot be used, then one alternative would be to use aligned fibres (which tend to give a higher level of properties such as shown in Table 4.8) or a thicker cross-section.

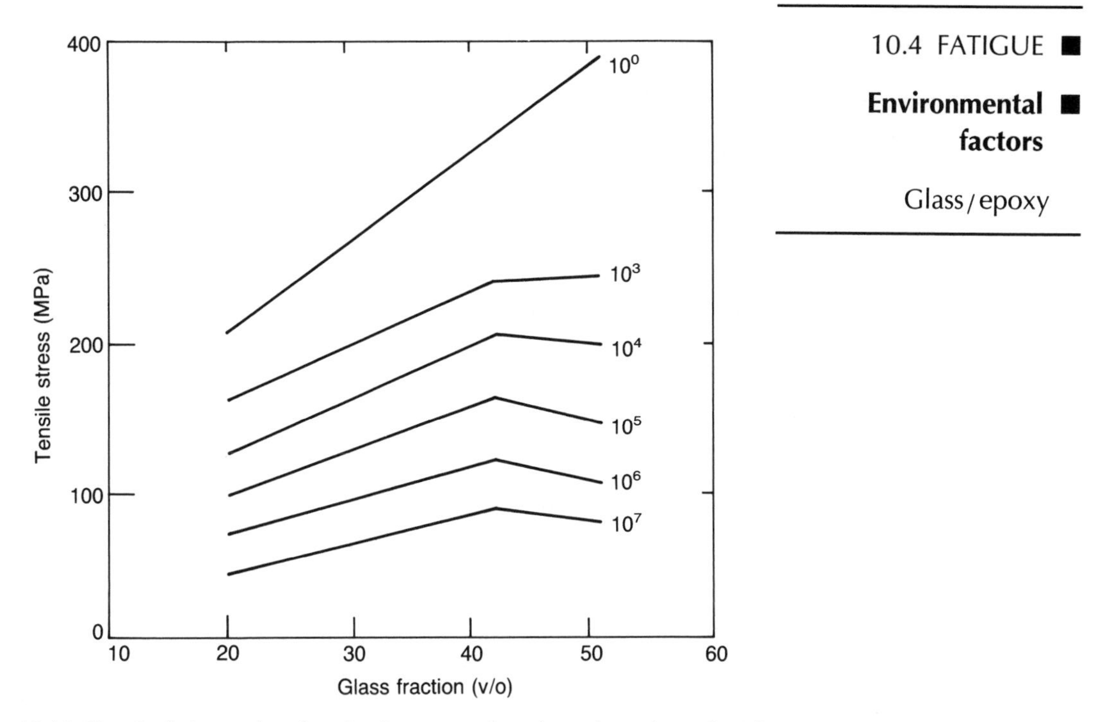

Figure 10.11 Tensile fatigue of a glass laminate as a function of number of cycles to failure (Creux).

Source

G. Creux at Vetrotex.

■ 10.4 FATIGUE

■ **Fibre alignment**

Influence of fibre alignment

The influence of fibre alignment is shown in Figure 10.12.

Materials

- reinforcement – random glass mat, V_f = 35% (Vetrotex); balanced warp/weft glass fabric, V_f = 45%; unidirectional E-glass rovings, V_f = 60% (Vetrotex); unidirectional R-glass rovings, V_f = 60% (Vetrotex);
- resin – polyester (mat), epoxy (fabric and UD).

Manufacture

Press moulding.

Testing

Tension, test frequency = 8.3 Hz, load control, minimum/maximum stress (R = 0.11).

Observations

This data set covers fibre alignment from random (mat) through orientated (balanced warp/weft) to fully aligned, the load always being applied along the highest fibre direction. The higher the alignment, the higher is the initial strength and the higher is the fatigue strength for a given number of cycles.

Because R-glass is stronger than E-glass (Table 3.1) the R-glass composite has a higher strength both initially and during fatigue; the strength difference between E- and R-glass composites decreases as the number of cycles increases.

Design implications

As with static strengths, the designer has to choose between higher strengths along fibre directions and lower strengths in other directions and more uniform properties but with lower values in specified or random directions.

Combination fabrics, which contain a mixture of fibre alignments, are often chosen to ensure that there are some fibres in all directions with more fibres in directions with higher loading.

10.4 FATIGUE ■

Fibre alignment ■

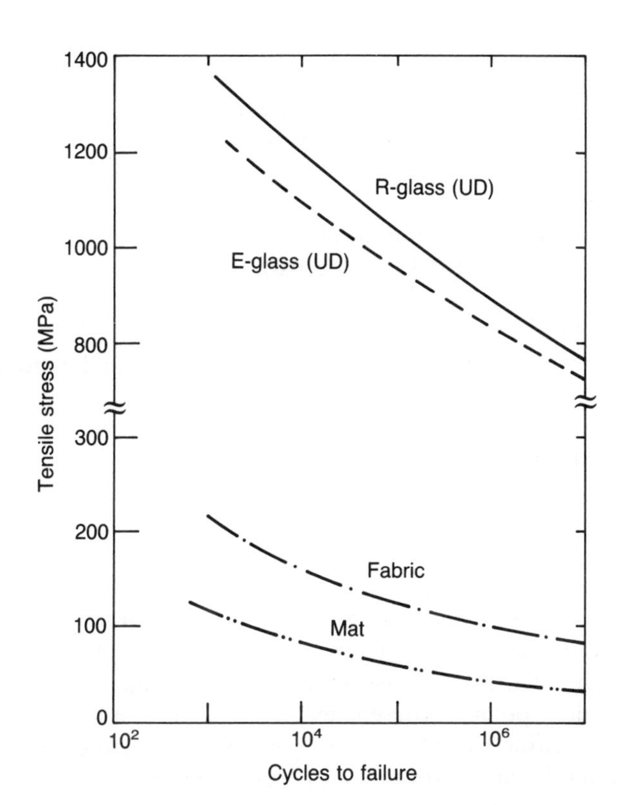

Figure 10.12 Effect of fibre alignment on the tensile fatigue of glass laminates; data are mid-point values obtained by Creux.

Source

G. Creux at Vetrotex.

■ 10.4 FATIGUE

■ **Fibre orientation**

Glass / polyester

Effect of fibre orientation

The effect of fibre orientation is shown in Figure 10.13 where rovings have been aligned at various angles to the load axis.

Materials

- reinforcement – continuous glass rovings;
- resin – polyester UP 333 (Hoechst).

Manufacture

Filament wound plates at various orientations to longitudinal axis; autoclave cured.

Test conditions

Constant amplitude, frequency 5 Hz, temperature 20°C, tensile loading ($R = 0.1$).

Observations

The data show how the damage accumulates at varying rates depending upon the fibre orientation. However at longer times there appears to be a tendency for the 0° and 10° samples to converge.

These data points represent extensive damage and effective failure; the loss in stiffness can be used to determine a suitable failure criterion. The normal failure strain provides a useful dimensionless parameter for characterizing fatigue.

Design implications

This type of fully aligned material represents the upper limit of what could be possible in a fibre-reinforced plastic as all the fibres are well aligned and the laminate manufactured under laboratory conditions. A lower level of properties would be obtained with less well aligned materials.

This work forms part of a very extensive investigation into the fatigue properties of materials funded by the R & D directorate of the CEC for the design of wind turbine blades.

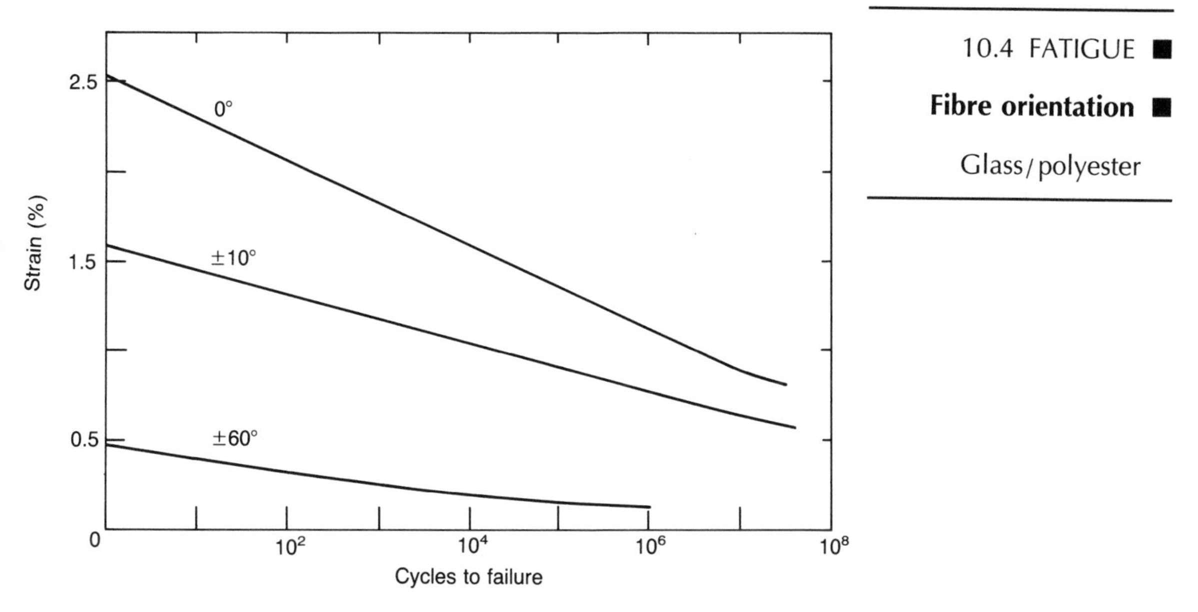

Figure 10.13 Effect of fibre orientation on the tensile fatigue performance of glass reinforced plastics (Lilholt and Andersen).

Sources

Lilholt, H. and Andersen, S.I. (1988) Fatigue behaviour of glass fibre reinforced polyester, *Proc. IEA Workshop on Fatigue in Wind Turbines*, ETSU, Harwell.

■ 10.4 FATIGUE

■ **Fibre length**

Resin type

Influence of fibre length

The influence of fibre length is illustrated in Figure 10.14 where composites with continuous rovings are compared with those with short length rovings (injection moulded components).

Materials

(a)

- reinforcement – continuous rovings E-glass Equerove EC 17 (Pilkingtons), R-glass P109 (Vetrotex);
- resin – epoxy MY750/HT972 (Ciba-Geigy).

(b)

- reinforcement – *c*.0.2 mm, E-glass;
- resins – nylon 66, polycarbonate, PPS, PAI.

Manufacture

(a) Filament winding of continuous rovings.
(b) Injection mouldings of thermoplastic compounds.

Test conditions

Load control, tensile loading (R = 0.1, 0.25).

Observations

The data bands for short length fibre thermoplastics and long fibre reinforced thermosets lie adjacent to one another with a similar shape if the fatigue data for each composite is divided by the ultimate strength (Table 10.3).

This normalization procedure allows not only the comparison of fibre

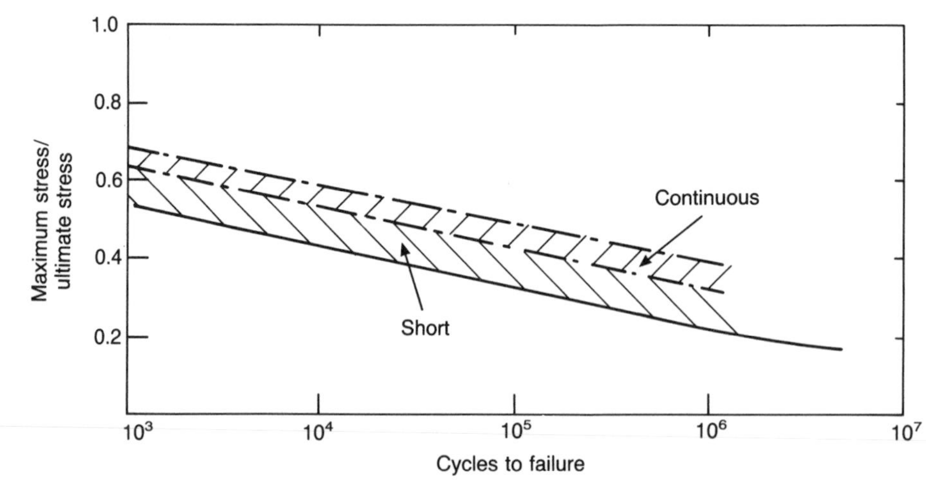

Figure 10.14 Tensile fatigue of both continuous and short length glass reinforced plastics (Mayer *et al.* and Mandell *et al.*).

Table 10.3 Initial ultimate tensile strengths of continuous and short fibre composites

fibre		continuous		short (0.2 mm)			
		R-glass	E-glass	E-glass			
resin		epoxy	epoxy	nylon 66	polycarbonate	PPS	PAI
UTS	MPa	1350	1050	260	200	180	200
V_f	v/o	67	64	24	24	24	24

lengths, but also resin type and glass type, and covers a wide range of ultimate strengths.

Design implications

These data imply that the fatigue strength is dominated by fibre type, i.e. glass rather than resin type or fibre length (provided the fibres are longer than the minimum length to transfer the load).

These data tend to verify the design rule for GRP of a decrease in fatigue strength of about 10–12% for each decade increase in the number of fatigue cycles up to one million cycles (see Figure 10.29 for a fuller discussion). There is evidence that at lower strains (and higher cycles) that this rate of decrease is reduced.

The normalization procedure has been applied to other design parameters like temperature (Figure 10.17), loading rate (Figure 10.20) and presence of holes and notches (Figure 10.22).

Sources

Mandell, J.F. *et al.* (1980) Fatigue of glass and carbon fiber reinforced engineering thermoplastics, *Proc. 35th Annual Technical Conference*, SPI, Paper 1, section 20D.

Mayer, R.M. *et al.* (1984) Some aspects of the fatigue behaviour of fibre reinforced thermosetting resins. *Composite Structures*, 2, 305–13.

■ 10.4 FATIGUE

■ **Resin types**

Glass fabric

Effect of resin type

The influence of resin type and blend is shown in Figure 10.15.

Materials

The *materials* are:

- glass – balanced warp/weft woven fabric 800 g/m^2, chopped strand mat 100 g/m^2 Rovimat (Chromarat);
- resins – ortho-polyester, Norpol 41-90 (Jotun Polymer); iso-polyester, Norpol 72-80 (Jotun Polymer); iso-NPG-polyester, Norpol 20-80 (Jotun Polymer); flexible vinyl ester, Norpol 92-20 (Jotun Polmyer); rubber-modified vinyl ester, Norpol 92-40 (Jotun Polymer).

Testing

Load control, tension/compression. Minimum/maximum stress ($R = -1$), temperature 20°C, frequency 2 – 5 Hz.

Observations

The data tend to fall into two groups. First, a high amplitude fatigue region where the effect of the matrix can be discerned. The NPG/ iso-polyester and the vinyl ester resins have a somewhat longer life for the same stress amplitude than iso- and ortho-polyesters. Secondly, a lower amplitude, higher cycle region where all the laminates have a similar lifetime.

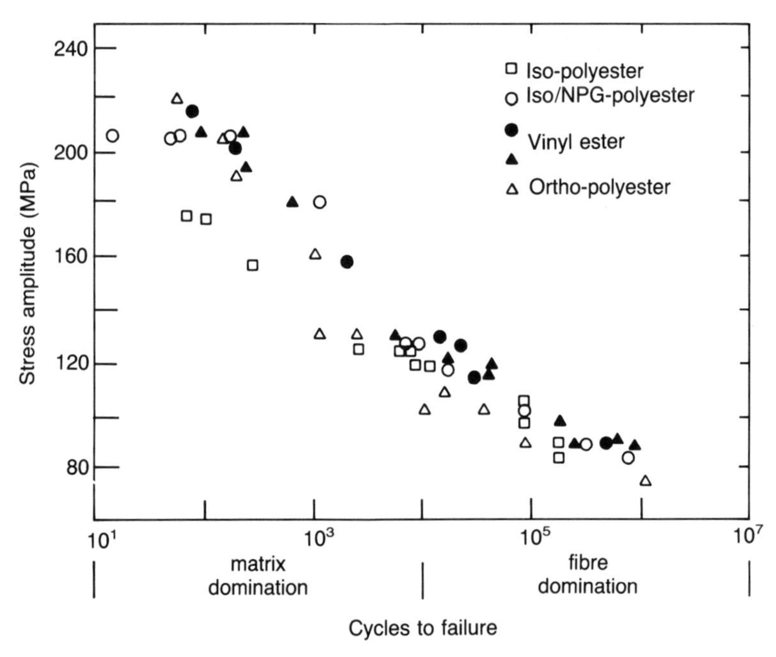

Figure 10.15 Tensile fatigue of different resin types and blends showing fibre and matrix dominated properties (Echtermeyer/SAMPE).

Design implications

The importance of stress amplitude during fatigue is quite marked for this type of reinforcement (combination glass fabric). The difference between matrix dominated and fibre dominated properties is determined by the presence of a knee point in a static load/deflection test (Figure 1.5(c)). The average loss in strength per decade of cycles approaches 10%, which is typical of glass-reinforced plastics.

Source

Echtermeyer, A.T. (1991) Significance of damage caused by fatigue on mechanical properties of composite laminates, *Proc. International Conference on Composite Materials 8*, Hawaii.

■ 10.4 FATIGUE

■ **Matrix type**

Glass/fabric

Influence of matrix

The performance under flexural fatigue of polyester and phenolic composites with the same reinforcement is shown in Figure 10.16.

Materials

- reinforcement – combination fabric, 600 g/m^2 woven roving, 200 g/m^2 chopped and needled glass (Fleming Laces) v/o 35%;
- resins – phenolic, J2027 (BPCL), polyester, 73.2661 (DSM).

Manufacture

Resin transfer moulding.

Test conditions

Span/depth 16:1, minimum/maximum stress, $R = 0.1$, frequency = 5 Hz, temperature 20°C.

Observations

There is a monotonic decrease in strength at a similar rate for both glass/polyester and glass/phenolic composites. This implies that the fatigue strength is being governed by the reinforcement rather than the matrix.

Design implications

Though the available data are limited, these results suggest that phenolic-based composites possess good fatigue resistance similar to glass/polyester provided that an appropriate fibre size has been selected. Coupled with its good high temperature performance, this makes it an attractive material for applications involving structural loading.

Further measurements would be required to validate any design.

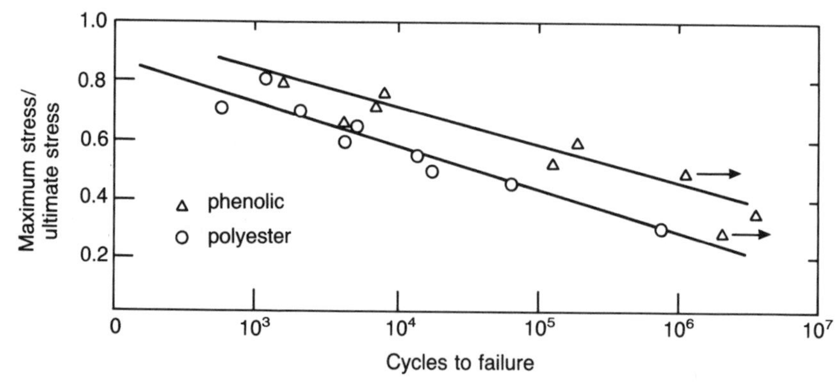

Figure 10.16 Flexural fatigue of phenolic and polyester laminates in 3-point bending tests (Forsdyke).

Source

Forsdyke, K.L. (1988) Phenolic FRP today, *Proc. British Plastics Federation Conference*, Blackpool.

Effect of temperature

Materials

- reinforcement – balanced warp/weft, fine weave E-glass fabric; V_f = 45%;
- resin – epoxy, Permaglass 22 FE (Fothergill and Harvey).

Manufacture

Hot pressed moulding.

Test conditions

Tensile, $R = 0.1$, various temperatures.

Observations

Glass thermosets can sustain higher fatigue loads at low temperatures than at high temperatures. This arises because at low temperatures the composite becomes stronger and stiffer, while at high temperatures the composite becomes weaker as the service temperature approaches the HDT of the resin.

The data can be reconciled by normalizing the data by the UTS measured at that test temperature. Figure 10.18 shows the good agreement between S–N curves for samples heated at various temperatures and hence shows the pattern by which fatigue strength increases with reduced temperature.

The increased scatter at the lowest stresses (and high cycles) could indicate a fatigue limit.

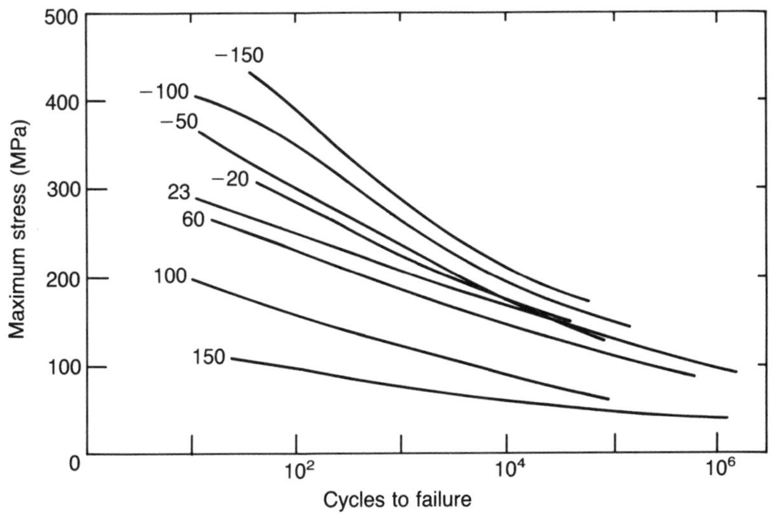

Figure 10.17 Comparison of S–N curves of glass/epoxy laminates loaded in tension at different test temperatures (Sims and Gladman).

■ 10.4 FATIGUE

■ Temperature

Glass / epoxy

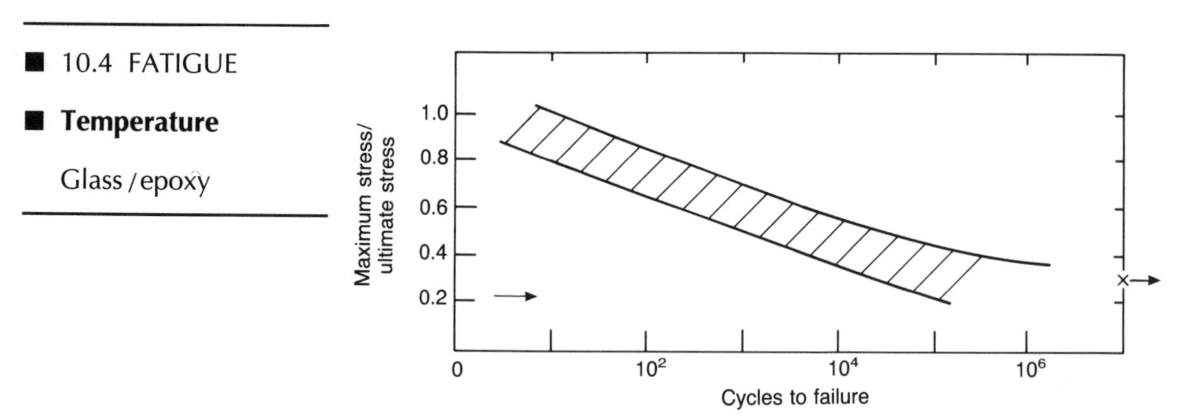

Figure 10.18 Normalized S–N curves of glass/epoxy laminates at different test temperatures (Sims and Gladman).

Design implications

A useful method for obtaining information on fatigue behaviour at various temperatures is to measure a fatigue curve at one temperature and to determine the static strength at any other temperature of interest.

By constructing a normalized curve like Figure 10.18, it should then be possible to predict the fatigue curves at other temperatures. To finalize the design it would be necessary to measure the fatigue resistance at the maximum service temperature required.

Source

Sims, G.D. and Gladman D.G. (1982) *A Framework for Specifying the Fatigue Performance of Glass Fibre Reinforced Plastic, Report DMA(A) 59*, National Physical Laboratory, Teddington.

Influence of reverse loading

Materials

- reinforcement – combination fabric; 700 g/m^2 UD, 100 g/m^2 chopped glass, powder bonded (Ahlstrom);
- resin – iso-polyester (DSM).

Manufacture

Contact moulding by hand lay-up; v/o = 35%.

Test conditions

Constant amplitude, tensile loading (R = 0.1); reverse (tensile-compressive) loading (R = −1).

Observations

For a given number of fatigue cycles the composite is able to sustain a higher working strain under tensile loading than under a combination of tensile/compressive loading. If a stiffness reduction of 10% is used as a failure criterion then the reduction in fatigue life is about half a decade in these materials when reverse loading is introduced.

Design implications

In only few applications is the loading tensile in nature and it is more likely to contain an element of flexural or reverse loading. So the loading mode does need to be assessed in the design as well as an appropriate failure criterion.

The bending of the fatigue curves towards the horizontal suggests that a fatigue limit might be present in these materials at low enough stresses. This could be an important design consideration if the component is to be subjected to prolonged endurance running. What little other evidence is available suggests that this limit could well be the fatigue limit of the matrix to micro-cracking.

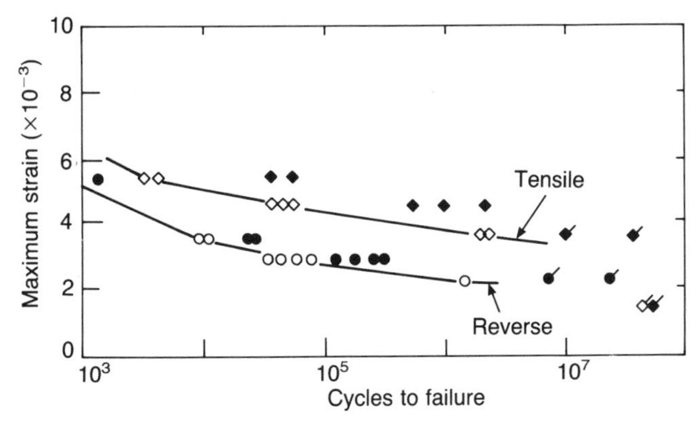

Figure 10.19 Fatigue behaviour of diagonally plied samples loaded in tension or reverse loading (Bach/ECN).

Source

Bach, P. (1988) High cycle fatigue of glass fibre reinforced polyester, *Proc. IEA Workshop on Fatigue in Wind Turbines*, ETSU, Harwell.

■ 10.4 FATIGUE

■ **Loading rate**

Glass/epoxy

Effect of loading rate

The effect of varying the loading rate is shown in Figure 10.20 for the same material.

Materials

- reinforcement – fine weave balanced glass fabric;
- resin – epoxy, Permaglass 22 FE (Fothergill and Harvey).

Manufacture

Hot pressed moulding of prepreg; $V_f = 45\%$.

Test conditions

Load control, tensile loading ($R = 0.1$); loading rate of 10^2 N/s to 10^6 N/s equivalent to test frequencies from 0.002 Hz to 50 Hz; temperature 20°C with temperature rise only at 50 Hz.

Observations

Increasing the loading rate increases the apparent fatigue strength in the manner shown in Figure 10.20. The frequency at the highest loading rate was sufficiently high that the measurements had to be corrected for self-heating induced in the coupon during testing. The ultimate tensile strength is also dependent upon strain rate with the highest strength being recorded at the highest rate (Table 10.4).

The data in Figure 10.20 can then be normalized by dividing by the appropriate UTS (Figure 10.21). This enables the data to fit a narrow band with only some divergence at high cycles for the highest loading rate.

Design implications

These results suggest the fatigue strength is characterized by a single common response for a given composite and that specimens degrade at a similar rate.

This provides the justification and a means for screening materials by testing at a higher frequency (or loading rate) than that encountered in practice. Notable examples of such applications include those of leaf springs and wind turbine blades.

If loads are applied suddenly the material response of composites is good because the ultimate strength increases proportionally to the loading rate for many composites (Figure 6.7).

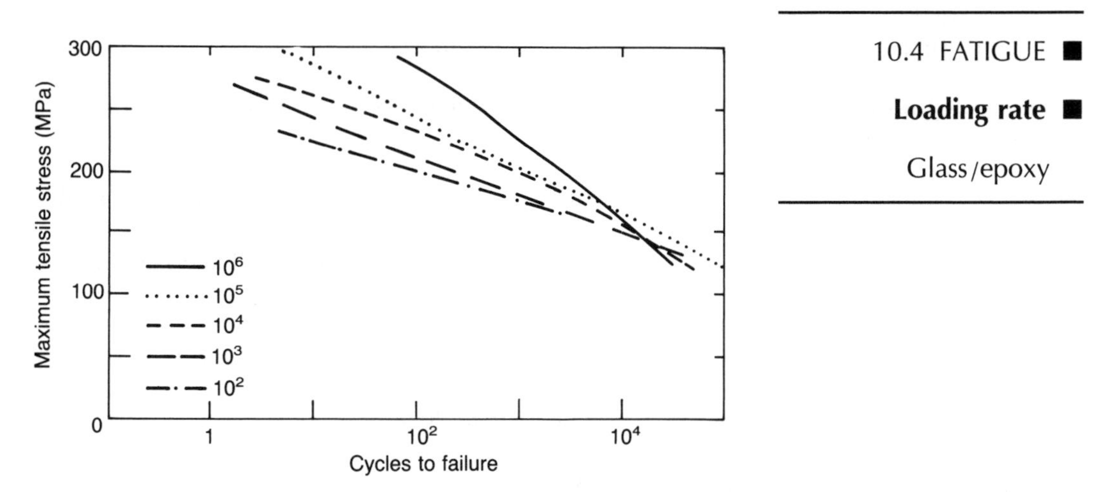

Figure 10.20 Effect of five different loading rates on fatigue life (Sims and Gladman).

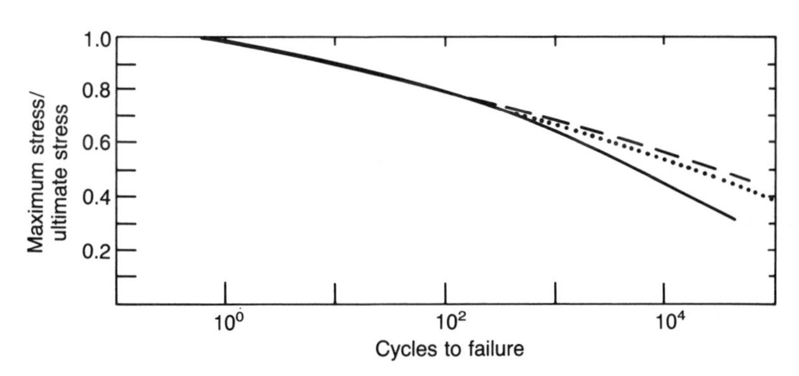

Figure 10.21 Effect of normalizing fatigue data at five different strain rates (Sims and Gladman).

Table 10.4 Effect of loading rate on UTS

loading rate	(N/s)	10^2	10^4	10^5	10^6
UTS	(MPa)	260	275	330	360

Source

Sims, G.D. and Gladman, D.G. (1978) Effect of test conditions on the fatigue strength of a glass fabric laminate – Part (A), frequency, *Plastics and Rubber: Materials and Applications*, **1**, 41–8.

■ 10.4 FATIGUE

■ **Holes and notches**

Glass / epoxy

Effect of holes and notches

The effect of holes and notches is shown in Figure 10.22.

Materials

- reinforcement – balanced warp/weft, fine weave E-glass fabric; V_f = 45%;
- resin – epoxy, Permaglass 22 FE (Fothergill and Harvey).

Specimens

Damage inserted as follows: central hole drilled 40% of specimen width; notches at both edges reducing specimen width by 50%; pre-loaded by a single cycle to 90% of mean UTS.

Test conditions

Tensile, R = 0.1, various temperatures.
10^5 N/s constant rate of loading.

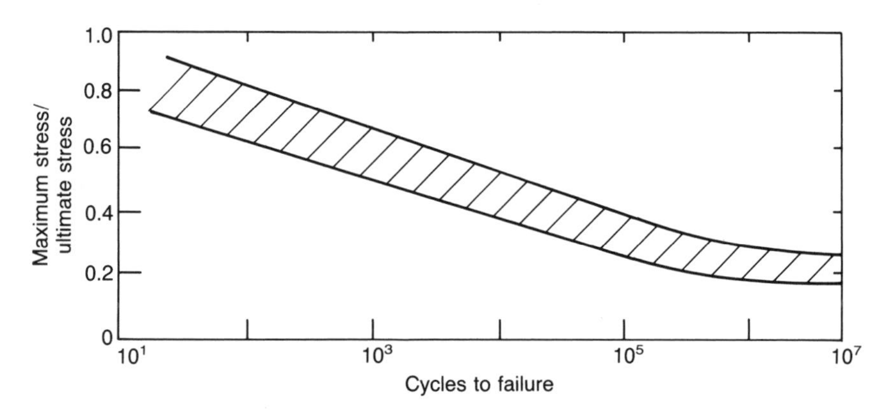

Figure 10.22 Fatigue curves for samples with various forms of holes and notches inserted before testing (Sims and Gladman).

Observations

By measuring both the fatigue curve and the UTS for each type of sample, the data could be normalized by dividing by the appropriate UTS to produce data which lie within a narrow band.

A similar methodology has been applied by the authors to account for:

- self-heating of samples during testing;
- specimen orientation;
- aggressive environments.

Design implications

This approach provides a framework for predicting and obtaining fatigue information for applications involving glass-reinforced plastics.

It enables material combinations to be rapidly screened for a wide range of properties.

Sufficient measurements need to be taken with the selected material combination to check that this method is valid.

Source

Sims, G.D. and Gladman, D.G. (1980) Effect of test conditions on the fatigue strength of glass fabric laminate – Part (B), Specimen condition, *Plastics and Rubber: Materials and Applications*, 3, 122–8.

■ 10.4 FATIGUE

■ **Fibre type**

Epoxy resin

Effect of fibre type

The influence of fibre type is illustrated in Figure 10.23.

Materials

- reinforcement – carbon HTS (Courtaulds), aramid Kevlar 49 (du Pont), E-glass fibre (Pilkingtons);
- resin – epoxy 69 (Fothergill and Harvey).

Manufacture

Prepreg material with balanced warp/weft; autoclave moulded; $V_f = 60\%$.

Test conditions

Load control, constant amplitude, constant rate of stress, tension.

Observations

The initial strain capability of glass/epoxy is greater than that of aramid/epoxy and carbon/epoxy. However, at long times the maximum working strain for all fibres appears to approach a level which is only marginally above the endurance limit of the matrix resin.

The decrease in strain is generally related to a reduction in strength rather than stiffness.

Design implications

If a high working strain is required, then allowance has to be made for the reduction in strain if glass and aramid fibres are used. Carbon, by

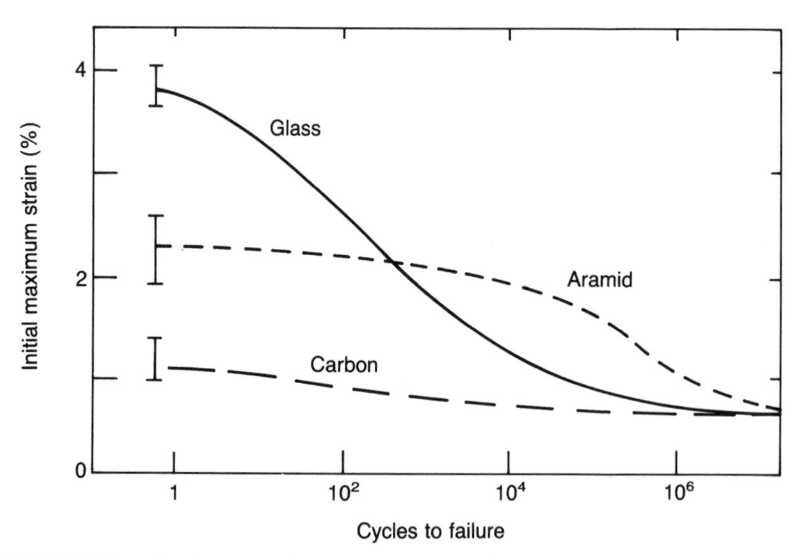

Figure 10.23 Tensile fatigue of prepreg moulded laminates with different types of reinforcement (Harris *et al.*).

Fibre type ■

Epoxy resin

virtue of its higher stiffness, reduces the strain in the matrix, so little damage accumulates even at high cycles.

Source

Harris, B. *et al.* (1990) Fatigue behaviour of carbon fibre reinforced plastics. *Composites*. **21**(3), 232–42.

■ 10.4 FATIGUE

■ **Moisture**

Epoxy resin

Effect of moisture

The effect of pre-conditioning coupons by drying, exposing to moisture or boiling in water is shown in Figure 10.24 for tensile fatigue.

Materials

- reinforcement – carbon HTS (Courtaulds), aramid, Kevlar 49 (du Pont), E-glass (Pilkingtons);
- resin – epoxy type 69 (Fothergill and Harvey).

Manufacture

Laminates moulded from UD prepregs with balanced warp/weft; pre-conditioning as per Figure 9.5.

Testing

Load control, constant amplitude, constant rate of stress, tension.

Observations

There are considerable differences between the fibre types. Carbon/epoxy is seemingly unaffected by any of the pre-conditioning treatments. Glass/epoxy is only affected by boiling and then for only cycles below 10^5. Aramid/epoxy has a complex response in that it is initially affected and then is relatively stable for some 1000 cycles before beginning a steady decrease. These fibres are unusual in that the dried composite appears to be less fatigue resistant than that containing moisture. The effect of pre-conditioning is illustrated in Figure 9.5.

Design implications

Pre-conditioning treatments produce varied responses and their effect might be greater on small coupons than on large panels. Carbon has the most favourable response, but may not possess the other balance of properties.

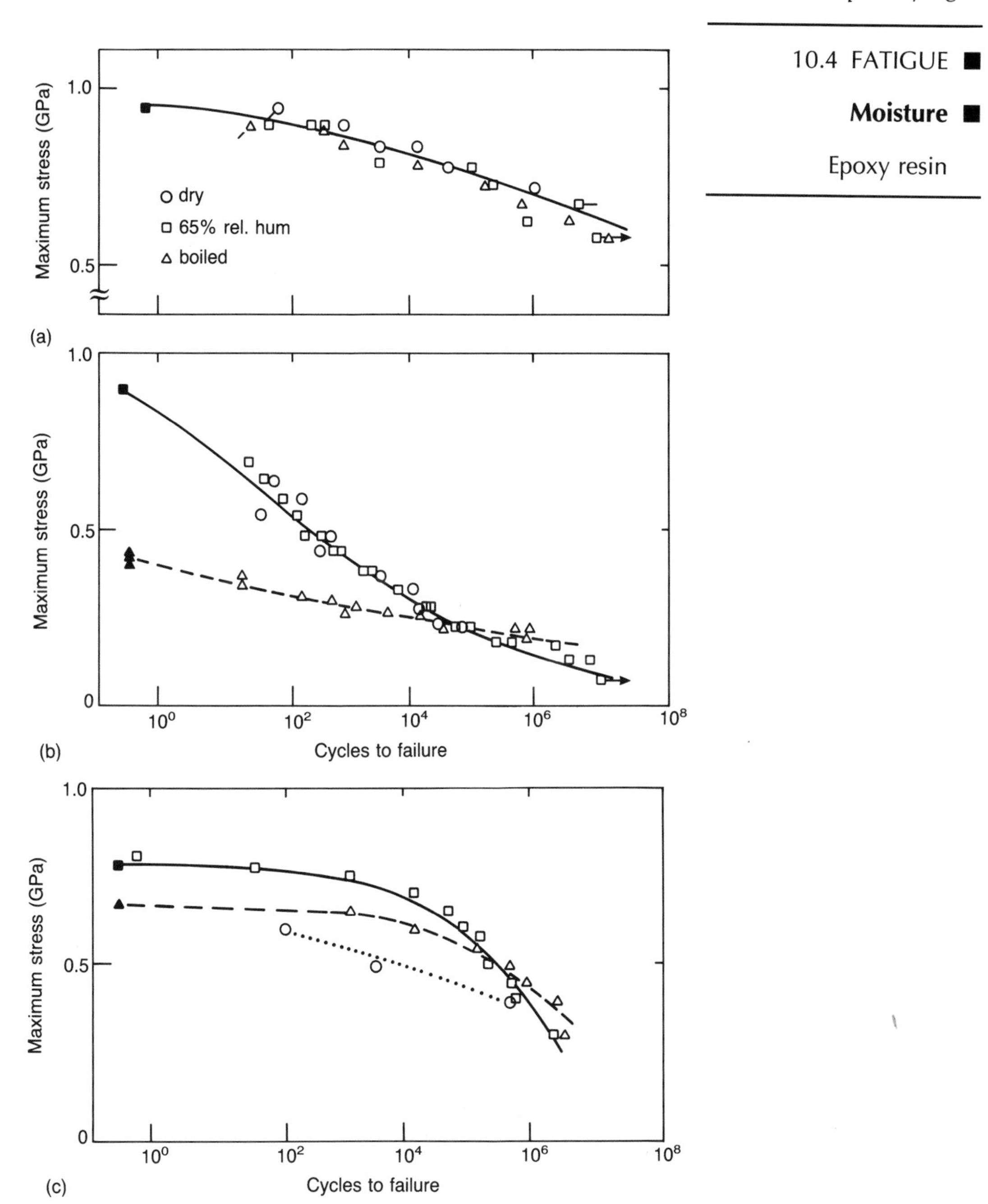

Figure 10.24 Effect of pre-conditioning on coupons undergoing tensile fatigue: (a) carbon/epoxy; (b) glass/epoxy; (c) aramid/epoxy; 65% rel. humidity for three months at 20°C (Jones *et al.*).

Source

Jones, C.J. *et al.* (1984) Environment fatigue behaviour of reinforced plastics. *Proc. Royal Society of London*, **396**, 315–38.

■ 10.4 FATIGUE

■ **Shear fatigue**

Epoxy resin

Shear fatigue

The shear fatigue of UD composites of carbon, glass and aramid fibres is shown in Figure 10.25.

Materials

- reinforcement – carbon fibre, Grafil HM-S (Courtaulds), carbon fibre, Grafil HT-S (Courtaulds), E-glass fibre (Silenka), aramid, Kevlar 49 (du Pont);
- resin – epoxy, MY 750/HY 906 (Ciba-Geigy).

Manufacture

Hand lay-up, 6 mm square rods, 60 v/o unidirectional fibres.

Testing

Frequency 0.17 Hz, torsional loading, temperature 20°C.

Observations

There is a rapid decrease in shear strain amplitude with fatigue cycles for all composites and typically 50% of the strain capability is lost between 10^3 and 10^4 cycles.

It is not clear from these data whether a fatigue limit does exist.

There are, however, significant differences in the torque/shear strain curves when tested statically. Whereas the aramid/epoxy has a linear stress/strain curve, those of carbon/epoxy and glass/epoxy are distinctly non-linear.

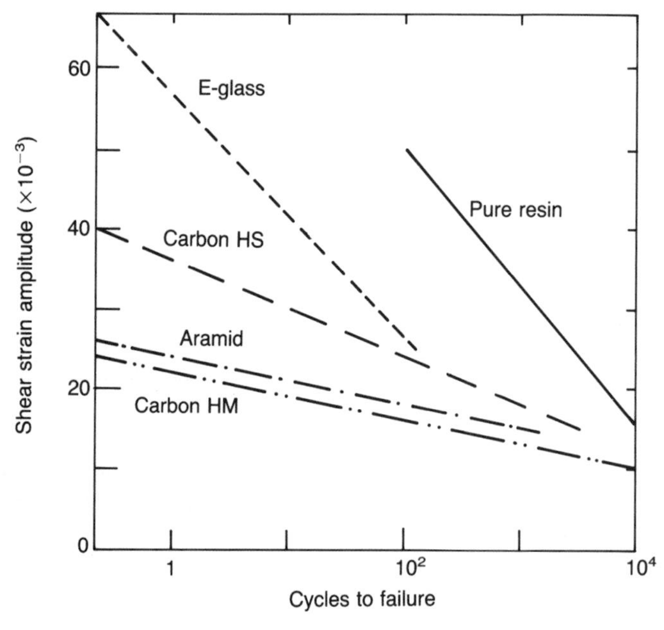

Figure 10.25 Variation of fatigue life with strain amplitude on torsional cycling (Philips and Scott).

Design implications

Cyclic shear stresses have severe effects on unidirectional fibre composites so it is important to keep shear strains low or ensure that fibres are placed in more than one direction.

These data were collected some years ago and fibre treatments/coatings are likely to have improved, thus improving strain capability of the fibre/matrix interface.

Fatigue of short beam interlaminar shear coupons or ±45° coupons does not result in such rapid fatigue failure.

Source

Philips, D.C. and Scott, J.M. (1977) Shear fatigue of unidirectional fibre composites. *Composites*. 8, 233–6.

■ 10.4 FATIGUE

■ **Compression**

Carbon/epoxy

Compressive fatigue

The effect of compressive loading is shown in Figure 10.26.

Materials

- reinforcement – carbon fibre, T300 (Toray);
- resin – epoxy, 934 (Ciba-Geigy).

Manufacture

Prepregs laid up at various fibre orientations, autoclave moulding, V_f = 62%.

Testing

Load control, compressive loading (R = 10), frequency = 1 Hz.

Observations

The need for a valid test is shown by the results for two different gauge lengths (15 and 10 mm). These represent a need to have a uniform stress state in the sample and yet not fail by Euler buckling.

These results show that for 0° aligned coupons there is only a slight reduction in the stress capability over seven decades of fatigue cycles *c*.15%.

The shear response of the ±45° aligned coupons to fatigue loading is somewhat greater, with a decrease of *c*.5% per decade of cyclic loading.

Design implications

These data have been collected on a material combination whose manufacturing route has been determined with care. As a result, the compressive and shear properties show good resistance to fatigue.

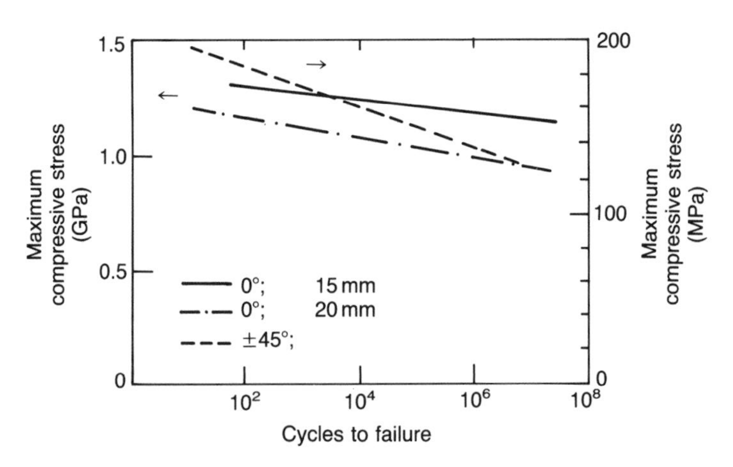

Figure 10.26 Compressive fatigue data for 0° and ±45° specimens of carbon/epoxy with different gauge lengths (Lifschitz/ASTM).

The importance of a valid compressive test should be checked by varying the gauge length.

Source

Lifschitz, J.M. (1988) Compressive fatigue and static properties of a UD graphite/epoxy composite. *Journal of Composite Technology and Research*, **10**(3), 100–6.

■ 10.4 FATIGUE

■ **Flexural stress**

Carbon fibre

Flexural fatigue of carbon/PEEK

Flexural testing of carbon fibres in PEEK and epoxy are contrasted in Figure 10.27.

Materials

- reinforcement – T300 carbon fibre (Toray);
- resins – PEEK, APC 2 (ICI), epoxy, Cycam 985 GT 6095-III (Short Bros.).

Manufacture

Compression moulding of prepregs at 0° (and other angles).

Testing

3-point flexural fatigue, $R = 0.1$, frequency = 5 Hz, temperature 20°C.

Observations

The flexural fatigue data for both resin composites show that carbon/PEEK has a lower degradation rate than that of carbon/epoxy for UD. The data for three alignments are contrasted in Table 10.5.

Carbon/epoxy is slightly better with cross-plied laminates (0/90°), than carbon/PEEK, but both have similar properties compared with angled-plied laminates (±45°).

Design implications

PEEK and epoxy have similar properties in flexural fatigue and it is from a consideration of other properties such as high temperature capability, processability or cost that specific advantages might accrue to one or other resin system.

The properties of PEEK composites are very dependent upon the crystallinity of the matrix material, so processing of such a thermoplastic must be undertaken with care.

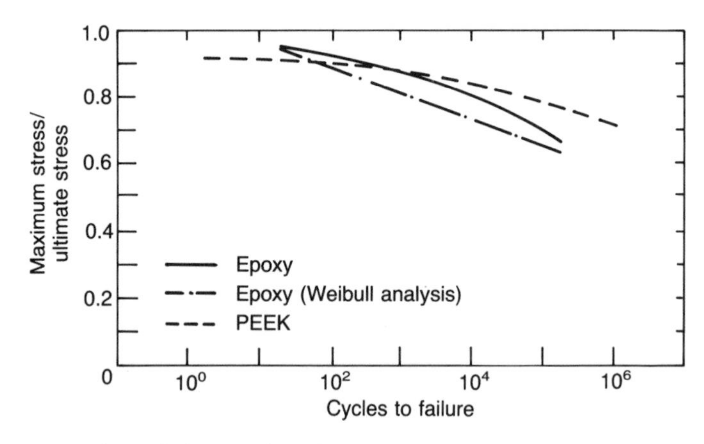

Figure 10.27 Flexural fatigue data for carbon/PEEK and carbon/epoxy; both for 0° laminates (Buggy and Dillon).

Table 10.5 Comparison of flexural strength of carbon fibre laminates and ratio of maximum fatigue stress to initial stress after 10^5 cycles

	PEEK	epoxy	PEEK	epoxy
alignment	UFS (GPa)		σ_m/σ_{ufs}	
0°	1.69	1.58	0.67	0.80
0/90°	1.24	1.02	0.80	0.85
±45°	0.80	0.72	0.65	0.65

Source

Buggy, M. and Dillon, G. (1991) Flexural fatigue of carbon fibre PEEK laminates. *Composites*, **22**, 191–8.

10.5 DESIGN STRATEGY

There are many parameters to consider in designing against creep and fatigue in composites, many of which have been illustrated in this chapter. A possible design strategy is set out in Figure 10.28.

Material selection

- select matrix capable of being processed according to desired fabrication route;
- select fibre type according to strain capability; check compatibility with resin as a good fibre/resin bond is essential;
- select fibre length according to stress level; long orientated fibres will provide continuity of load path and ensure principal loads resisted, which is deemed a prerequisite to withstand any combination of high temperature and stress;
- ensure fibre/matrix combination can be fabricated by selected route;
- like any strategy, if condition cannot be met, procedure must be iterated.

Design

- ensure that both strains and stresses are within acceptable limits;
- check properties if relevant data not available;
- select joint type and location to be compatible with temperature and stress level deemed acceptable;

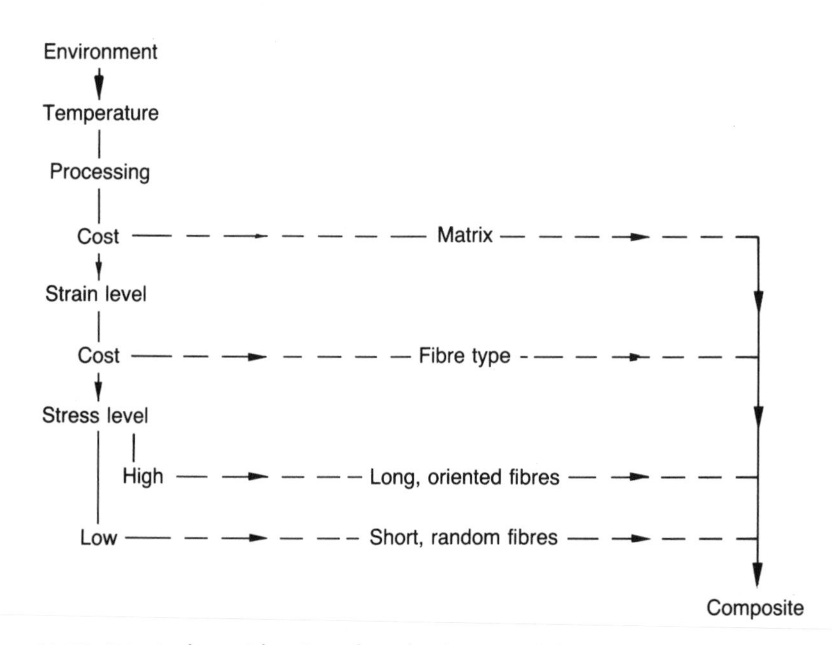

Figure 10.28 Principal considerations for selecting materials.

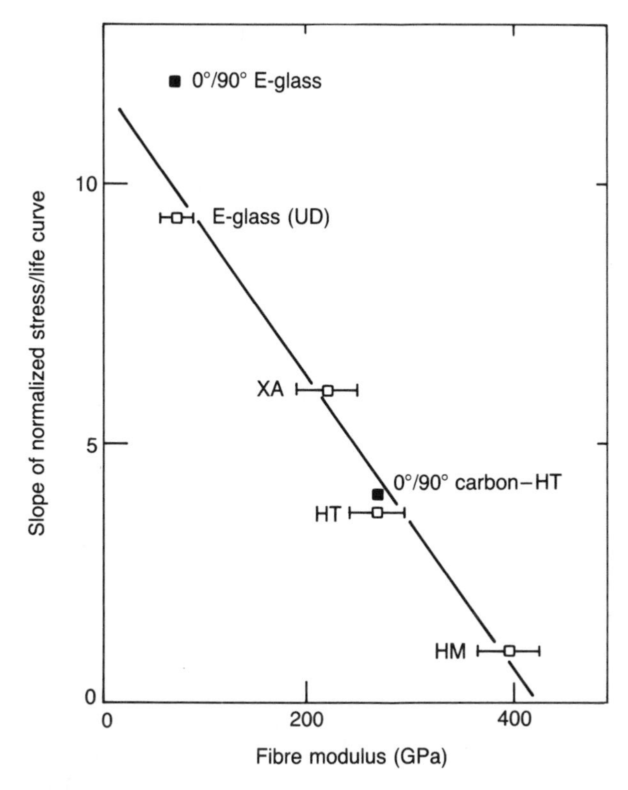

Figure 10.29 Slope of normalized stress/life curve against fibre modulus for fibre reinforced plastics. XA, HT and HM are carbon fibres with various tensile moduli (as in Figure 3.1) (Jones *et al.*).

- if possible, the design should ensure that stage III (Figure 10.1) is never reached (like Figure 10.7);
- if stage III cannot be avoided then ensure that degradation and damage accumulation can be detected;
- if degradation cannot be detected then if failure does occur it will be benign and not catastrophic.

The slope of the normalized stress/fatigue life curve versus fibre modulus is summarized in Figure 10.29. This shows the effect of fibre type, modulus and lay-up for carbon and glass composites. It is clear that the higher the modulus, the lower is the rate of loss of strength during fatigue.

Source

Jones, C.J. *et al.* (1984) Environment fatigue behaviour of reinforced plastics. *Proc. Royal Society of London*, **396**, 315–38.

10.6 REFERENCE INFORMATION

General references

Lilholt, H. and Talreja, R. (eds) (1982) *Fatigue and Creep of Composite Materials*, Risø National Laboratory, Roskilde – many papers are still of relevance.

Reifsnider, K.L. (ed.) (1991) *Fatigue of Composites*, Elsevier, Barking – in-depth survey.

Proc. IEA Workshop on Fatigue in Wind Turbines (1988) ETSU, Harwell – good on high cycle fatigue of glass composites.

Plastics for Aerospace Vehicles, Part 1: Reinforced Plastics, TSL, Hitchin – USA military specification MIL-HDBK-17B.

Creep

Holloway, L. (ed.) (1989) *Polymers and Polymer Composites*, Thomas Telford, London – good discussion on creep.

Other isolated papers in various conference proceedings.

Fatigue

Konur, O. and Matthews, F.L. (1989) Effect of the properties of the constituents on the fatigue performance of composites: a review. *Composites*, **20**, 317–28.

Work being undertaken into the fatigue and design of wing blades by a European consortia of institutes will be published in book form in 1994; for further information contact coordinator Rayner M. Mayer, Sciotech, 9 Heathwood Close, Yateley, Hampshire GU17 7TP, UK.

Search strategy

The RAPRA bibliograph base is the most convenient manner of identifying who has done what. Copies of the source documents are also available from RAPRA (refer to section 2.8).

Existing databases are not very helpful for these topics.

11 INFLUENCE OF PROCESSING ON PROPERTIES

SUMMARY

The major processes and their influence on important composite properties are considered. Since processing also limits the shape and size of the component, this needs to be decided at an early stage in the design process. Some data are collated from preceding chapters and some from studies carried out to identify the characteristic properties of processes.

11.1 INTRODUCTION

Processing has an important influence on properties by virtue of processing requirements which affect not only size and shape of a component, but also the orientation, length and size of the reinforcement.

The major processes are listed in Table 11.1; injection moulding is not considered further because of the restriction in this book to long length fibres.

Only contact moulding by hand lay-up does not require dedicated equipment and it is for this reason that this process is so popular.

Table 11.1 Major fabrication processes

type	abbreviation	comment
contact moulding	COM	hand lay-up or spray moulding; small production runs
resin transfer moulding	RTM	medium production; dimensionally accurate parts
press moulding	PRM	high speed moulding in long production runs
prepreg moulding	PPM	high consistency and high performance
pultrusion	PUL	longitudinally reinforced parts (for example, rods)
filament winding	FIL	bodies of revolution such as pipes and tanks
centrifugal moulding	CEM	generally for 3D shapes
injection moulding	IM	generally for small parts with short fibres (<5 mm)

The major process limitations are summarized in Table 11.2. Of these the maximum size is typical rather than the absolute value whilst shape, reinforcement length and orientation are limited by the type of process.

It is the combination of fibre length, orientation and ability to compress the fibre/resin mixture whilst moulding which determine the level of property achievable. Only some materials can be used with some processes and this is illustrated in Table 11.3.

Each of the processes is discussed in turn and some characteristic data are presented or cross-referenced from other chapters.

Table 11.2 Process limitations

process	COM	RTM	PRM	PPM	PUL	FIL	CEM
maximum dimension (m)	30	3	3	10	50	30	6
shape	any	any	flat	any	const	cyl	cyl
reinforcement length	any	any	any	cont	cont	cont	short
orientation	any	any	ori	ori	ori	ori	ran

Notes: Shape
- flat: height much less than in-plane dimensions;
- const: constant cross section;
- cyl: cylindrical symmetry.

Length
- cont: continuous;
- any: short, long or continuous.

Orientation
- ori: orientated;
- ran: random.

Table 11.3 Comparison of materials and processes (OCF)

	Thermosets					Thermoplastics										
	polyester	polyester moulding compounds	phenolic	epoxy	methacrylate	acetal	nylon 6	nylon 6/6	polycarbonate	polypropylene	polyphenylene sulphide	ABS	polyphenylene oxide	polystyrene	polyester PBT	polyester PET
injection moulding	●	●	●	●	●	●	●	●	●	●	●	●	●	●	●	
hand lay-up	●		●	●	●											
spray-up	●		●	●												
press moulding	●	●	●	●							●					
prepreg moulding	●		●	●												
filament winding	●		●	●												●
pultrusion	●		●	●	●											
resin transfer moulding	●		●	●	●									●	●	●
reinforced reaction injection moulding	●		●	●	●		●									

11.2 CONTACT MOULDING

This is the most versatile manufacturing process due to its ability to make almost any shape or size of component and to use a wide variety of fabrics.

A clear distinction needs to be made between the process of hand lay-up and that of spray-up as these give rise to different levels of properties (Table 11.4). The differences arise out of the processes themselves – spray-up being semi-automated in that pre-impregnated resin is sprayed onto a mould whereas hand lay-up requires individual layers to be laid up by hand and impregnated with resin.

The attainable properties are contrasted in Figure 11.1 for the two processes both having random orientations of chopped rovings. Hand lay-up has a further advantage in that the level of properties can be increased systematically by suitable selection of fabric rather than random mat as illustrated in Figure 11.2.

Table 11.4 Comparison of hand lay-up and spray-up

aspect	hand lay-up	spray-up
reinforcement length	any	short
reinforcement orientation	any	random
maximum fibre fraction	high	low
strength range	large	low
production rate	low	medium

■ 11.2 CONTACT MOULDING

■ **Processing**

Effect of processing

Materials

- spray-up – glass chopping rovings;
- hand lay-up – chopped strand mat (Vetrotex M4 450);
- resin – ortho-polyester (Jotun 44M);

Manufacture

Either spray or hand lay-up.

Observations

Both reinforcements provide a random orientation, the mat resulting in more fibres parallel to the mould surface. Consolidation is easier with the mat, but care is required with complex surfaces to ensure that rovings as well as resin are present.

Design implications

Hand lay-up produces a higher level of properties than spray-up for a given fibre fraction. However, since spray-up is semi-automated, it is quicker for laying fibres, particularly for large surfaces which might be lightly loaded.

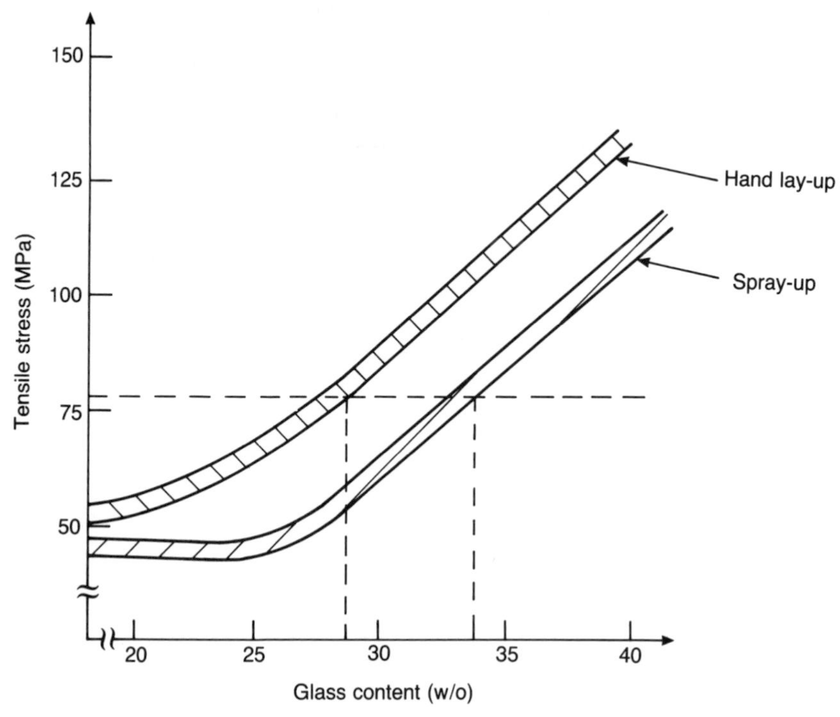

Figure 11.1 Effect of processing on tensile strength of composites (Arvesen).

Source

R. Arvesen at Jotun Polymer.

11.2 CONTACT MOULDING

Fabrics

Effect of fabric

Materials

- glass – chopped strand mat (<40 w/o), combination fabric (40–57 w/o), combination fabric plus UD fabric (57–65 w/o);
- resin – iso-NPG polyester (Jotun 20).

Manufacture

Contact moulding by hand lay-up.

Observations

The effect of substituting a balanced fabric for a random mat, and then adding some highly orientated fabric, is to increase the strength of the laminate in the direction of the greatest fibre orientation, and to decrease the strength in other directions. Other data are set out in Table 4.3.

Design implications

A wide range of property levels can be obtained by hand lay-up using suitable fabrics. This enables properties to be tailored to design requirements. The level of properties in other directions could then be calculated from the rule of mixtures and fibre orientation.

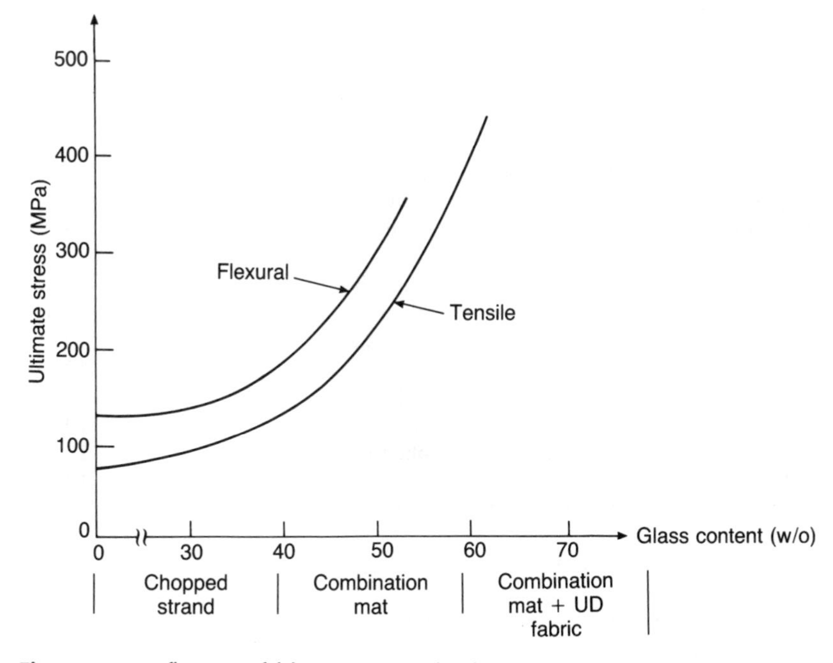

Figure 11.2 Influence of fabric on strength of composites (Arvesen).

Source

R. Arvesen at Jotun Polymer.

■ 11.2 CONTACT MOULDING

■ **Properties**

Collated data

The principal data from other chapters are listed in Table 11.5 for reference.

Summary

The principal advantages of contact moulding are listed in Table 11.6.

Table 11.5 Principal data for hand lay-up

property	glass reinforcement	resin	reference (Table)	reference (Figure)
strength, modulus	fabric	vinyl ester	4.6	
		polyester	4.6	
		phenolic	4.8	
		epoxy	4.12	
	mat	thermoplastic	4.14	
shear	fabric, mat	polyester	4.4	
impact	mat	polyester	6.1	6.4
flammability	mat	phenolic	8.4	
		methacrylate	8.5	
smoke generation	mat	phenolic	8.8	8.6
		polyester		8.6
		epoxy		8.6
		methacrylate	8.8	8.7
water absorption strength	UD fabric	polyester		9.3
moisture uptake	all fibre types	epoxy	9.5	9.5
		polyester	9.5	
corrosion resistance	fabric	many resins		9.8
weathering	fabric	phenolic	9.8	
stress rupture	mat	phenolic		10.5
		polyester		10.5
creep strain	fabric	polyester		10.7
tensile fatigue	fabric	polyester		10.15
		vinyl ester		10.15
flexural fatigue	fabric	phenolic		10.16
		polyester		10.16

Table 11.6 Principal considerations for using contact moulding

advantages	disadvantages
all reinforcements	labour intensive
all resins	long manufacturing time
wide level of properties	reproducibility dependent on skilled labour
cheap tooling	

11.3 RESIN TRANSFER MOULDING

This process is almost as versatile as contact moulding as regards selection of resin and reinforcement, and therefore the possible range of properties. Its primary advantage is that of reproducibility and reduced labour, though this is offset by increased tooling cost as it is a closed mould technique.

Because the resin is injected into a mould containing the fabric, the key need is for good wet-out between the resin and the rovings as the resin needs to dissolve the binder (finish) which holds the rovings in a bundle.

Most fibres are designated as epoxy or polyester-compatible, but if other resins are to be used then the material suppliers need to be consulted as to the appropriate combination of fibre and resin.

The best way of checking compatibility is to compare the properties of the fibre/resin combination with a known combination either from this book or by measurement.

Pre-forming

A thermoplastic or thermoset binder can be used to bind layers of fabric or fibre bundles together. This allows specific orientations to be achieved and placing of local reinforcement as required by the design.

On softening the binder (*c*.50°C), the fabric can be deformed to any sensible shape by use of light pressure. Fibre bundles can be aligned by a spray-up process or by coating with a binder before being sprayed onto a former (Figure 11.3). A chopping gun could be used to produce chopped rather than continuous fibres for better shape. On cooling, the shape is maintained because the binder hardens. This process is called 'pre-forming'.

The great advantage of this process is that the fibre distribution can be checked before the resin is added, ensuring that an appropriate level of properties can be maintained throughout the component.

No deleterious effect of binder concentrations in the range of 2–8% on the properties has been detected.

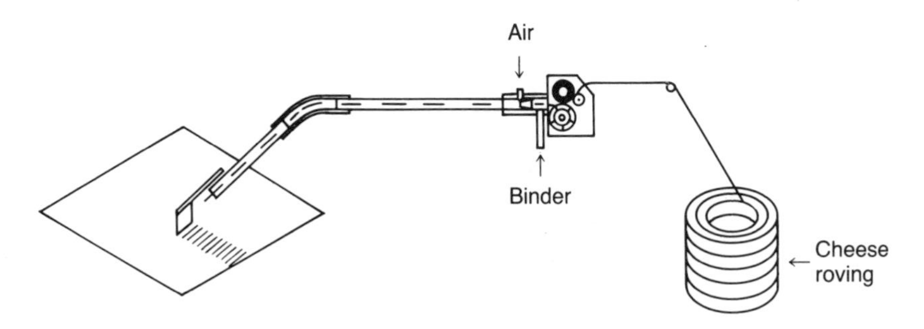

Figure 11.3 Orientated continuous fibres drawn from a roving cheese, being coated with a binder and laid down in a pre-form shape (OCF).

■ 11.3 RESIN TRANSFER MOULDING

■ **Preforming**

Property level

Provided the reinforcement is distributed as specified and there is adequate wet-out between fibre and resin, property levels are similar to that obtainable with hand lay-up. As the RTM process uses a closed mould, all surfaces can be moulded to size and provided with a good surface layer for protection.

Source

Jander, M. (1991) *Industrial RTM: New Developments in Moulding and Preforming Technologies*, Technical report, Owens Corning Fiberglas, Battice.

Collated data

The principal data from other chapters are listed in Table 11.7 for reference.

Summary

The principal advantages of RTM moulding are listed in Table 11.8.

Table 11.7 Principal data for RTM

property	glass reinforcement	resin	reference (Figure)
shear strength	CSM, WR, UD rovings	phenolic	4.12
flexural	combination	phenolic	10.16
fatigue	fabric	polyester	10.16

Table 11.8 Principal considerations for using RTM

advantages	disadvantages
most reinforcements	wet-out of fibre and resin
most resins	design of mould
wide level of properties	medium tooling cost
reproducibility	

■ 11.4 PRESS MOULDING

■ **Properties**

11.4 PRESS MOULDING

Press moulding is used to cover a range of processes in which a mould is filled with material comprising fibre/resin/filler. The mould is then closed and pressure (and usually temperature) is used to cure the mixture thereby moulding the component.

The moulding compounds separate into those using thermosetting resins (SMC, BMC, DMC) and thermoplastics (TSC, GMT) and are used with specific processes (Table 11.9). These can contain either random, chopped or orientated glass fibres (Table 2.7). As compounds are developed for specific properties and applications, both the process route and property level are known and will be reproducible.

Collated data

The principal data from other chapters are listed in Table 11.10 for reference.

Typical data

Some typical data are given in Table 11.11 for two compounds and processes.

Observations

A higher fibre volume fraction can be obtained with SMC than BMC compounds, which is reflected in the higher value of properties for the former than the latter. However the compounds come in different forms (sheet and dough for SMC and BMC, respectively) and this dictates the shape and geometry of the mould.

Specific properties can be tailored for a given process by altering the formulation of the compound and the process conditions. Two principal concerns are maintaining the initial fibre length and ensuring that the flow into the mould does not produce areas which are deficient in fibres.

Table 11.9 Types of press moulding

moulding process	pressure	temperature	materials
cold press	yes	no	various reinforcements thermosetting resins
hot press	yes	yes	moulding compounds SMC, DMC, TSC, GMT
compression-injection	relaxed, then applied	yes	BMC

Table 11.10 Principal data for press moulding

property	reinforcement	resin, compound	reference (Table)	reference (Figure)
strength, modulus	short fibre	SMC, DMC GMT, TSC	2.8	
	glass mat	GMT	4.14	
impact	UD fabric	epoxy	6.1	
toxic gas emission	short fibre	phenolic SMC	8.8	
weathering	short fibre	SMC		9.19
fatigue, temperature	balanced fabric	epoxy		10.17, 10.18
fatigue, loading rate	balanced fabric	epoxy		10.20
fatigue, holes	balanced fabric	epoxy		10.22

Table 11.11 Typical data for moulded BMC and SMC components

property	SMC*	SMC*	BMC*	BMC†
fibre fraction (w/o)	30	50	22	22
tensile strength (MPa)	80	160	41	33
tensile modulus (GPa)	12	16	12	10
tensile elongation (%)	<1.0	1.7	0.5	0.5
flexural strength (MPa)	179	310	88	87
flexural modulus (GPa)	11	14	10	10
compressive strength (MPa)	165	220	140	
impact strength Izod (J/m)	860	1047	230	160

* Compression moulding.
† Injection moulding.

Design implications

As the level of properties is only one consideration for design, both compounds have been successfully used in volume production to make outer body panels for vehicles.

Source

A Comparison of Materials and Processes for Fiber Glass Composites (1986) Owens Corning Fiberglas, Toledo.

Demez, K. (1990) *Pièces Automobile Moulées par Injection de BMC*, Owens Corning Fiberglas, Battice.

■ 11.4 PRESS MOULDING

■ **SMC and RTM**

Use of SMC and RTM in cars

The requirements for the Espace included the production of *c*.40 units per day. As this number tended to fall between those of various GRP processes, Renault established the properties and unit costs for SMC moulding and RTM moulding (Figure 11.4, Table 11.12).

Observations

There is little difference between the two sets of properties at room temperature though differences are more apparent at 110°C. Both use different materials and processes. Considerable development work in both material formulation and press closure was undertaken to achieve a high quality finish.

Design implications

The cost per unit part dictated the choice of moulding process as the properties are similar. The graph indicates that as the production quantity increases, the SMC process first at low pressure and then at high pressure becomes more cost effective.

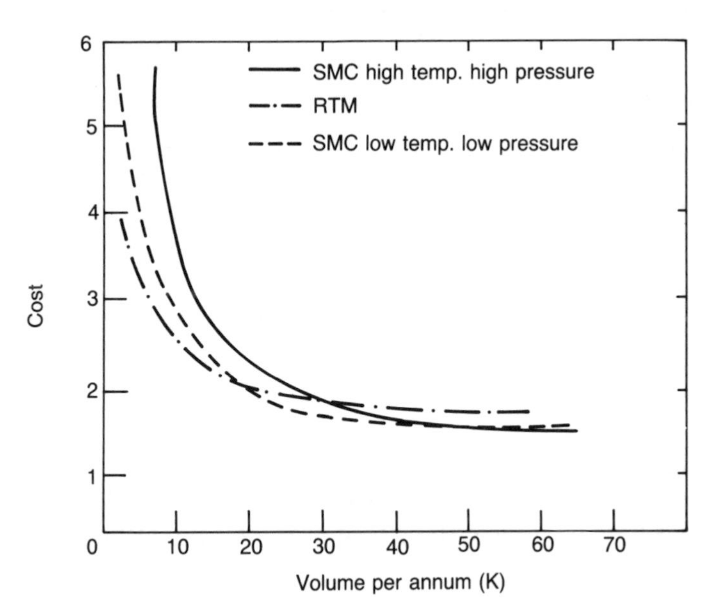

Figure 11.4 Variation in component price in terms of production quantity and moulding technique (Gurliat and Ollive/Menzolit).

Table 11.12 Flexural properties of polyester SMC and RTM mouldings

property	SMC	RTM
strength at 23°C (MPa)	120	140
strength at 110°C (MPa)	70	40
modulus at 23°C (GPa)	8	7
modulus at 110°C (GPa)	6	4
impact strength Charpy (kJ/m^2)	42	40
shrinkage (%)	+0.08	0.2

Source

Gurliat, A. and Ollive, C. (1988) Low pressure and temperature moulded SMC parts for the Renault Espace car – four years of results. *Engineering Plastics*, **1**(4).

■ 11.4 PRESS MOULDING

■ **BMC**

Compression-injection moulding of rear door

Fiat had a requrement for a rear hatch-back door for the UNO Turbo capable of being produced at the rate of several hundred per day. Other requirements included using the existing hinges of the UNO itself, component stiffness and external surface appearance which matched that of the steel panels. The design concept included incorporation of a rear spoiler and a double wall section between two skins. The material properties for the two skins are listed in Table 11.13.

Observations

The material properties of the two skins are very similar to one another. However the compound formulations differ considerably in order to achieve the design requirements such as surface finish.

The low value of properties compared with say that of the SMC formulation for the Espace can be accommodated by making use of two skins and a double wall cavity in between.

Design implications

The choice of a moulding process has to be related to both the production quantity and the geometry of the component. This component has a complex profile so that it could be better filled, without fibre degradation, through a two-step procedure (injection followed by compression), especially when changes of thickness range between 2 and 4.

As with SMC production of the Renault Espace, the appropriate formulation of the compound and level of quality control has been the key to producing a successful product.

Table 11.13 Mechanical properties of moulded BMC panels

property	inner skin	outer skin
glass content (w/o)	20	20
shrinkage (%)	0.05	0.05
flexural strength 20°C (MPa)	9.5	7.3
flexural strength 120°C (MPa)		5.6
flexural modulus 20°C (GPa)	0.9	0.8
impact strength Charpy (kJ/m^2)	30	25

Source

Rossi, F. *et al.* (1987) *Thermosetting Composite Injection and Compression Technology for a Double Wall Rear Door of a Passenger Car*, ANTEC.

Summary

The principal reasons for using press moulding are given in Table 11.14.

Table 11.14 Principal considerations for using press moulding

advantages	disadvantages
resin and reinforcement pre-selected	restricted range of materials
materials formulated to optimize properties	high tooling cost
properties known	high capital cost
suitable for medium to high volume production	

■ 11.5 PREPREG MOULDING

■ **Properties**

11.5 PREPREG MOULDING

The advantage of this process is that the reinforcement (in the form of long fibres) is already impregnated with a partially cured resin. Either vacuum or light pressure (in an autoclave) is required to consolidate layers during curing. The cure schedule for optimum properties will have already been determined by the material supplier.

There is a close analogy to press moulding of long fibres, in the sense that the materials are pre-selected, though the temperatures and pressures are lower.

Prepreg materials are thin, typically 0.2 mm thick, so that many layers are required to obtain a reasonable thickness, which does permit layers to be orientated so as to obtain specified properties. Drape can be a problem because of the continuous nature of the fibre so care is required in laying up.

Collated data

The principal data from other chapters are listed in Table 11.15.

Summary

The principal reasons for using prepreg moulding are listed in Table 11.16.

Table 11.15 Principal data for prepreg moulding

property	reinforcement	resin	reference (Table)	reference (Figure)
strength, modulus	glass (E and R)	epoxy polyester	5.2	5.1
effect of temperature	E-glass	epoxy phenolic bismaleimide PPS PEEK	5.4	5.5 5.5 5.5 5.5 5.5
strength, modulus	aramid	epoxy	5.7	
fibre modulus		epoxy	5.9	
volume fraction		epoxy	5.10	5.6
shear strength		epoxy	5.11	
effect of temperature			5.12	
strength, modulus	carbon	epoxy PAS PPS PEEK	5.14 5.22	

Properties

Table 11.15 *Continued*

property	reinforcement	resin	reference (Table)	reference (Figure)
fibre modulus	carbon	epoxy bismaleimide	5.16, 5.17 5.18	
transverse strength		epoxy	5.19	
effect of temperature		epoxy bismaleimide polyimide PEEK	5.21	
impact	carbon	epoxy polyimide PES PEI PEEK	6.2	
fracture energy	various	epoxy PEEK	6.3	
thermal	various	epoxy PEEK phenolic	7.3, 7.4	
electrical	various	epoxy	7.8	
smoke generation, flash-over	various	epoxy phenolic PEEK		8.8, 8.9
moisture	various	epoxy	9.2, 9.3	9.5
fatigue, loading rate	glass	epoxy		10.20
fatigue, environment	various	epoxy		10.24

Note: 'Various' refers to glass, aramid and carbon fibres. The designer should check availability of materials with suppliers.

Table 11.16 Principal considerations for using prepreg moulding

advantages	disadvantages
resins and reinforcements prescribed	restricted range of materials
process route established	drapability limited
property level known	thin layers
economic in small volumes	high material and labour cost

■ 11.6 PULTRUSION

■ **Mechanical properties**

Glass / vinyl / ester

11.6 PULTRUSION

Pultrusion is a process which involves pulling resin-impregnated rovings and/or fabrics through one or more heated dies to provide the profile shape and cure the composite. A wide variety of shapes can be produced.

Because of the continuous nature of manufacture, process conditions have to be optimized; thus good and reproducible components and properties can be attained. Good fibre alignment is achieved through complete automation of the reinforcement lay-up. A variety of fibre orientations can be achieved by using a suitable combination of rovings and fabrics.

Typical data

There is a considerable amount of data available principally from a few suppliers and moulders. Consequently, designers are able to incorporate the information directly into the design provided that a suitable profile and fibre orientation/property level are available.

As very little mechanical data have been quoted in the text, some typical properties follow.

Properties of profiles

Properties in the form of data sheets are available from some moulders. A set of properties for a structural profile is given in Table 11.17.

Materials

- reinforcement – E-glass random mat, UD rovings, polyester surface veil;
- resin – fire retardant vinyl ester;
- structural section – box, angle and I beam.

The comparative performance of two profiles, pultruded GRP and steel, is listed in Table 11.18.

Table 11.17 Mechanical properties of a pultruded structural section

property		longitudinal	transverse
tensile strength	(MPa)	300	55
tensile modulus	(GPa)	17	7
flexural strength	(MPa)	300	100
flexural modulus	(GPa)	12	5
compressive strength	(MPa)	200	100
compressive modulus	(GPa)	17	7
shear strength	(MPa)	60	60
shear modulus	(GPa)	3	3
Poisson's ratio		0.22	0.1
fibre fraction	(v/o)	50	

Table 11.18 Properties of pultruded and steel sections

property		box section		angle section	
		steel	GRP	steel	GRP
tensile strength	(MPa)	265	300	265	300
tensile modulus	(GPa)	210	19	210	19
maximum bending moment	(kN.m)	2.4	3.5	1.0	1.1
flexural rigidity	($kN.m^2$)	44	5.6	28	3
mass	(kg/m)	4.4	1.2	4.7	0.95
density	(Mg/m^3)	7.8	1.65	7.8	1.65

Profile sections

- steel box – 50 × 50 × 3 mm;
- GRP box – 50 × 50 × 4 mm;
- steel angle – 50 × 50 × 6 mm;
- GRP angle – 50 × 50 × 5.75 mm.

Observations

The data in Table 11.17 indicate the mean values for both longitudinal and transverse properties. The balance could be altered to suit specific requirements. Because the process is reproducible, both the standard deviation and the coefficient of variation could be readily determined.

The box and angle sections in Table 11.18 are for similar sizes and wall thicknesses. It is apparent that whereas the strengths are similar for the two materials, the stiffness is dissimilar, which reflects the difference in modulus between steel and glass fibre.

Design implications

Pultruded sections are attractive alternatives to steel sections where either strength or mass is important. If extra stiffness is required of the GRP sections, this could be achieved by increasing the overall dimensions.

Source

Quinn, J.A. (1988) *Design Manual of Engineered Composite Profiles*, Fibreforce, Runcorn.

■ 11.6 PULTRUSION

■ **Reinforcement type**

Glass / polyester

Effect of reinforcement type

The effect of varying the reinforcement from unidirectional rovings through to balanced combination fabrics is shown in Figure 11.5 and Table 11.19.

Materials

- reinforcement – (a) unidirectional rovings, (b) unidirectional plus combination fabric, (c) combination fabric;
- resin – iso-polyester, low shrinkage (Norpol 60/06, Jotun Polymer).

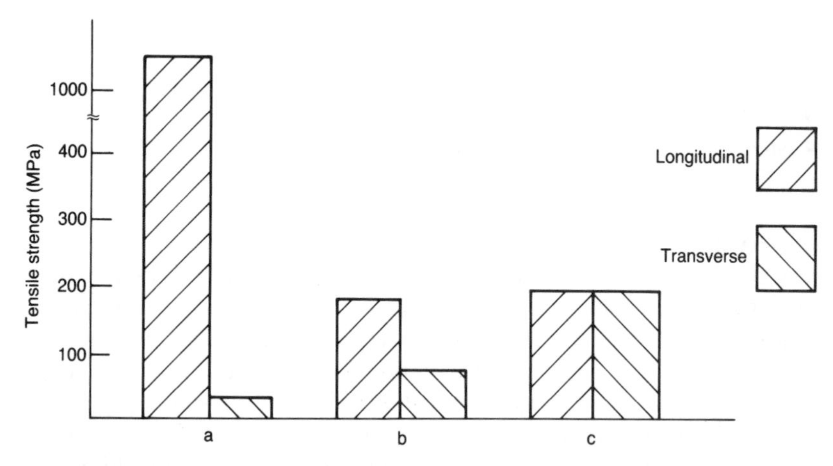

Figure 11.5 Variation in longitudinal and transverse tensile strength for three different types of reinforcement whose fibre orientation varies from (a) unidirectional; through to (c) a balanced warp and weft (Arvesen).

Table 11.19 Effect of different types of reinforcement in pultruded profiles

property		direction	reinforcement (a)	(b)	(c)
fibre fraction	v/o		75	65	65
tensile strength	(MPa)	long	1000	375	200
		trans	40	75	200
tensile modulus	(GPa)	long	50	27	
		trans		9	
strain to failure	(%)	long		1.7	1.8
		trans		1.5	
flexural strength	(MPa)	long	1000	375	275
		trans		150	275
flexural modulus	(GPa)	long	50	17	*c.*15
		trans		11	*c.*15

Table 11.20 Principal considerations for using pultrusions

advantages	disadvantages
materials pre-selected	not all material combinations possible
property level established	only certain types of profile possible
reproducibility very good	special profiles have a high tooling cost
profiles available off-shelf	
small or large numbers possible	

Observations

The principal effect of varying the reinforcement is to alter the ratio of the longitudinal to transverse properties. However, the more reinforcement in the transverse direction, the less in the longitudinal direction for a given fibre fraction and vice versa.

Design implications

To select a suitable profile, it may be necessary to determine the ratio of properties and therefore specify the required properties in the two principal directions.

If stiffness is important in the longitudinal direction then the majority of the fibres need to be located in this direction or the outer dimensions of the profile increased.

Summary

The principal reasons for using pultrusion are listed in Table 11.20.

Source

R. Arvesen at Jotun Polymer.

■ 11.7 FILAMENT WINDING

■ **Properties**

11.7 FILAMENT WINDING

Filament winding involves winding rovings onto a mandrel. The rovings may be impregnated with resin prior to being wound onto the mandrel. For thermoplastic resins, the rovings are pre-impregnated with resin by the supplier and the resin needs to be softened prior to winding onto the mandrel.

As a consequence of the process, roving alignment is known so that reproducibility is very good. If fibres are required along the length of the axis, a tape can be used rather than rovings in which the rovings are transverse to the length of the tape.

The process is generally used for components with cylindrical symmetry like pipes. A typical construction is shown in Figure 11.6.

Collated data

The principal data from other chapters are listed in Table 11.21.

Creep and cyclic fatigue of pipes

The response of filament wound pipes to static and dynamic loading by internal pressure is shown in Figure 11.7.

Materials

- reinforcement – glass rovings;
- resin – epoxy.

Manufacture

Filament winding ±55°C. Lay-up as shown in Figure 11.6.

Testing

ASTM D 2992

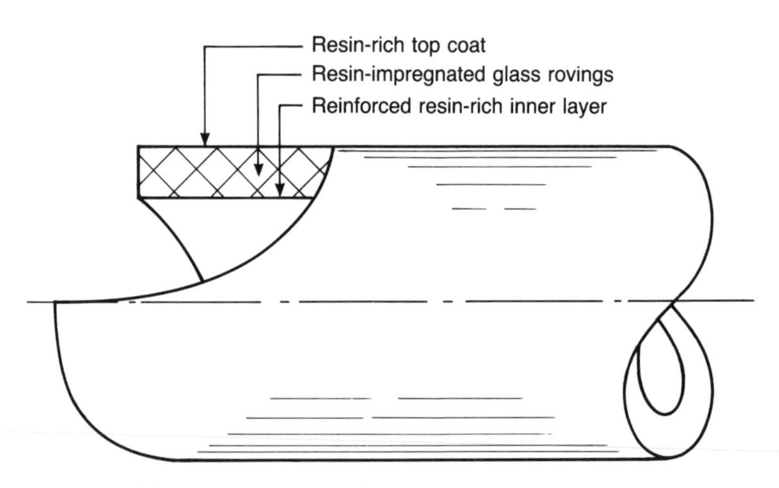

Figure 11.6 Typical lay-up construction for a GRP pipe (Hooning).

11.7 FILAMENT WINDING

Properties

Table 11.21 Principal data for filament winding

property	reinforcement	resin	reference (Table)	reference (Figure)
strength, modulus	glass	phenolic	4.8	
effect of temperature	glass, carbon	phenolic		4.13
	glass	polyester		4.14
		epoxy		
		vinyl ester		
		phenolic		
shear	aramid	epoxy	5.11	
creep	glass	vinyl ester		10.9
fatigue	E-, R-glass	epoxy		10.14
fibre orientation	glass	epoxy		10.13

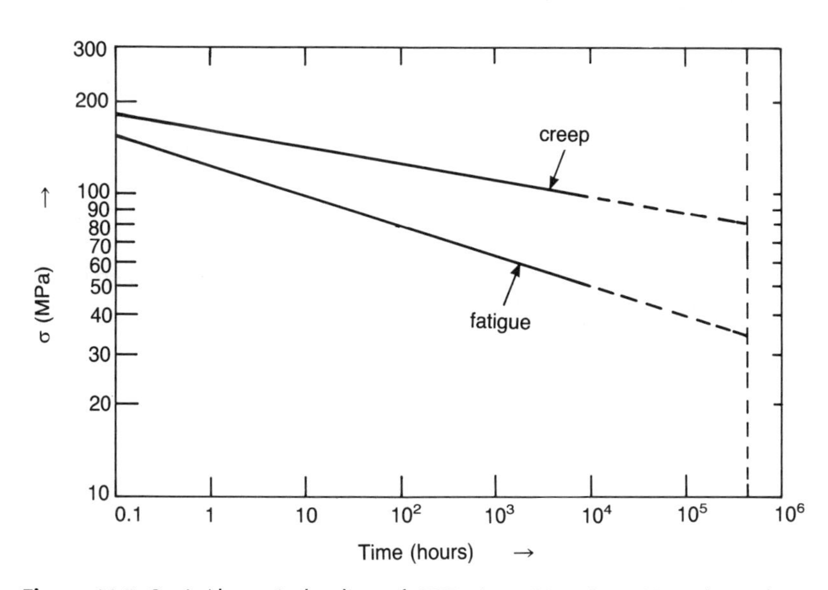

Figure 11.7 Static/dynamic loading of GRP pipes (Hooning). Note that 1 hour is the equivalent of 1500 fatigue cycles.

Observations

The hydrostatic strength decreases monotonically with time and is around 90 MPa after 10,000 hours. Cyclic loading has a greater effect and a strength of 50 MPa would be appropriate.

■ 11.7 FILAMENT WINDING

■ **Properties**

Table 11.22 Principal considerations for using filament winding

advantages	disadvantages
all fibres and most resins	restricted geometry unless multi-axis winding head used
good fibre alignment	
reproducible properties	
medium volume production	medium capital cost
easily automated	

Design implications

For pressure loading of pipes, measurements are essential to determine the long-term strength retention as well as the likely mode of failure. Extrapolation by a factor of 10 is permissible under certain conditions (section 2.6) in order to establish the design allowable stress.

This pipe system has been shown to provide excellent chemical resistance against many corrosive fluids encountered in the chemical process industry.

Summary

The principal reasons for using filament winding are set out in Table 11.22.

Source

Hooning, F.H. Growing interest in high pressure, large diameter GRP pipe lines, in *Proc. Conference in Plastics Pipes VI.*

11.8 RECYCLING

With the increasing use of plastic and reinforced plastics, a number of projects are under way to recycle such materials.

From BMC and SMC parts used for vehicle components, two types of material have been produced:

- a fibre fraction;
- a powder fraction containing fibre, which can be used as a filler in either SMC or BMC compounds.

Some data for the latter are shown in Table 11.23.

It is apparent that the process is commercially viable and that will encourage the recycling of other reinforced thermosets.

For thermoplastic composites, the recycling of glass mat thermoplastics has been investigated. One possibility is to remelt GMT mould directly into new shapes.

An alternative approach is to process GMT into short glass/thermoplastic pellets for injection moulding. Enhancement of the resulting properties lies between that of talc and (virgin) glass reinforcement (Figure 2.2).

Table 11.23 Properties of BMC laminates containing 29% recycled material

comparison with standard material	advantages	disadvantages
mass reduction	5%	
loss of impact strength		10%
loss of flexural modulus		5%
improvement in micro-waviness	25%	
increase in cost per kg		5%
cost per unit volume	identical	

Sources

Manducher and Cie (1991) Recycling polyester thermosets in Class A applications, in *French Composites Conference*.

BASF tackles 'difficult' plastics recycling cases, *Plastics and Rubber Weekly*, 28 July 1990.

11.9 DESIGN STRATEGY

A design strategy for selecting a suitable process is outlined in Figure 11.8.

Some process limitations are listed in Table 11.2 and materials used with specific processes in Table 11.3. Other information is listed in the summary table of each process and the collated data.

It is likely that the design loop listed above may have to be iterated several times possibly at each design stage. For a fuller description of processes and design strategies, consult Mayer (1993).

To deal with recycling, the constituent materials should be chosen during the initial design stage with reference to its re-use function. Defining the ultimate aim of recycling will help to optimize the choice of process and cost.

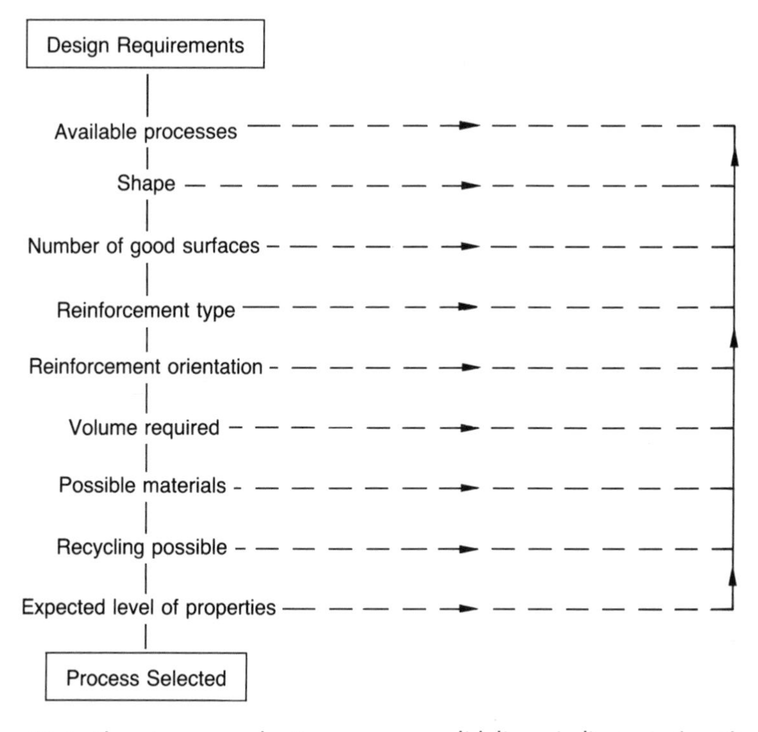

Figure 11.8 Choosing a production process: solid lines indicate 'go' options and dotted lines 'stop and think' options.

11.10 REFERENCE INFORMATION

Manufacture

There are several books which deal with various aspects of manufacture.

Mayer, R.M. (ed.) (1993) *Design with Reinforced Plastics*, Design Council, London – Chapter 6.
Richardson, T. (1987) *Composites – A Design Guide*, Industrial Press Inc, New York – Chapter 4.
Engineered Materials Handbook, Volume 1: Composites (1987) ASM International, Metals Park, Ohio – section 8.

Owens Corning Fibreglas, Toledo, Ohio have produced two good pamphlets:

Designing with fibre glass composites
A comparison of materials and processes for fibre glass composites

For pultruded products, *The Design Manual of Engineered Composite Profiles* by J.A. Quinn (Fibreforce, Runcorn) is invaluable. MMFG (Chatfield, Minnesota) have developed load/deflection tables for fibreglass profiles used for floor gratings.

Recycling

This is rapidly advancing topic because of political and social pressure to recycle waste in general.

Bouvier, J.M. *et al.* (1987) Managing fibreglass reinforced polyester composite wastes. *Resources and Conservation*, 5, 299–308 – an early review of the topic.
Europe to recycle auto thermoset waste 'this year', *Plastics and Rubber Weekly*, 4 May 1991 – sets out plans of ERCOM project – a joint European initiative.
Falkenstein, G. (1991) Recycling von Verbundwerkstoffen, AVK-Tagung 24, Berlin.
Jutte, R.B. and Graham, W.D. (1991) *Recycling SMC*, Owens Corning Fiberglas, Toledo – good set of properties.
BMW plans to recycle UK bumpers, *Plastic and Rubber Weekly*, 6 July 1991 – material to be granulated and then reprocessed into floor mats, boot liners and other components.
Blancon, R. (1991) Le recyclage des composites en Europe et aux Etas-Unis. *Composites*, 3, 179–82.
Desnost, C. (1991) Plastics materials recycling zero waste concept. *Composites*, 3, 177–8.

12 QUALITY ASSURANCE

SUMMARY

The requirements for quality assurance are considered in the light of properties only being defined as components are manufactured. Checks on incoming materials, process control, destructive and non-destructive evaluation are discussed and for each stage the use of data is considered.

12.1 INTRODUCTION

Fibre-reinforced plastics form a class of materials whose properties are only attained as the component is moulded. Any strategy for quality assurance must therefore take this primary factor into account.

The ISO 9000/9003 set of standards provides a framework for managing quality control systems and is widely used throughout the world. Some of the key requirements are listed in Table 12.1.

These aspects are considered in turn and related where possible to data presented in previous chapters.

Table 12.1 Quality control

aspects
incoming materials
processing
non-destructive evaluation
destructive testing

12.2 INCOMING MATERIALS

Materials should be purchased from suppliers to an agreed specification. It is good practice to check an incoming batch by undertaking a random test of one or more properties, depending upon the nature of the application.

Some possible properties are listed in Table 12.2, and relate to:

- fibres – tex from mass of a length of roving (section 3.4), amount of size by burn off (for glass fibres only), strength – tensile test of individual fibres;
- fabric – mass of a standard size say 300 × 300 mm, type and method of construction (section 3.4), number of roving ends (warp and weft), size content by burn off (glass only);
- resin – viscosity, which tends to increase as resin ages, gel time when standard amount of catalyst, accelerator added, Barcol hardness when resin fully cured;
- accelerator catalyst – gel time when added in standard amounts to resin; note if gel time high, it could either be resin, accelerator or catalyst;
- filler – particle size;
- powder binder – softening point.

Compatibility of constituents

It is prudent to make a test laminate by the selected fabrication route to check the compatibility of the individual constituents. Properties like interlaminar shear, strength, modulus, should be measured and compared with one or more of the following:

- relevant data in this text;
- manufacturers' recommendations;
- literature;
- previous measurements.

An outside test house can be utilized if no internal facilities are available.

Table 12.2 Possible checks of individual constituents

constituent	possible properties	other properties	reference
fibre	colour	size	
	texture	strength	Table 3.1
fabric	mass	size	
	type		Figures 3.7–3.8
resin	viscosity	gel time	Tables 4.1, 4.5, 4.9
		Barcol hardness	Figure 12.5
accelerator, catalyst	gel time		
filler	colour	particle size	Table 2.6
powder binder	colour	particle size	

■ 12.2 QUALITY CONTROL

■ **Components**

Barcol hardness

Barcol hardness

This is a micro-hardness technique, which uses a hand-held indentor to determine the hardness of a resin and hence the degree of cure of a composite. The relationship to strength was discussed in section 9.4.

The indentor is used at random over a plate or component and a histogram of the results should be plotted (Figure 12.1).

The requirement for averaging a number of indentations is two-fold:

- to ensure that an area is checked;
- to be able to eliminate the effect of any pore or hole that is encountered.

For a laminate or component, the indentation is sufficiently small (*c*.0.1 mm) that the indentor may strike either a resin or fibre rich area giving a value appropriate to either. For that reason, the spread in values is always greater for a laminate than for a pure resin.

The Barcol hardness of a component (Figure 12.2) is shown in Figure 12.3. The relationship of hardness to strength for various laminates has already been discussed (Figure 9.15). If a laminate has a very thick gel coat or surface layer then the indentor will only give a value appropriate to the surface (as with any surface technique).

Traceability

Materials need to be traceable along the supply chain from the supplier until they are combined and formed into a component. Suitable documentation is required and identification/batch numbers retained and recorded together with processing and NDT results on batch/item cards.

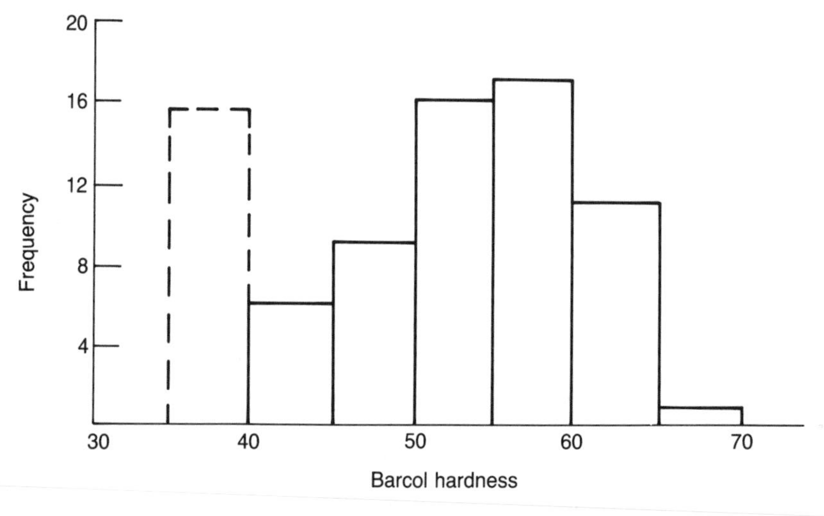

Figure 12.1 Barcol hardness histogram. The dotted column is for a cast resin; the others are for a needle mat fabric / phenolic resin (Mayer).

12.2 QUALITY CONTROL

Components

Barcol hardness

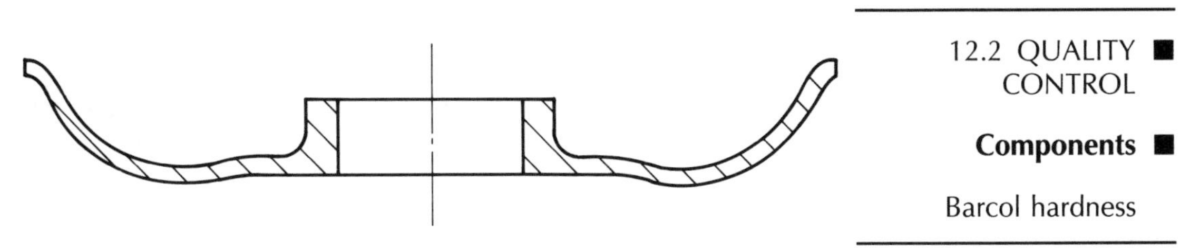

Figure 12.2 Cross-section of a moulded component; diameter is 490 mm, wall thickness is 6 mm (Mayer).

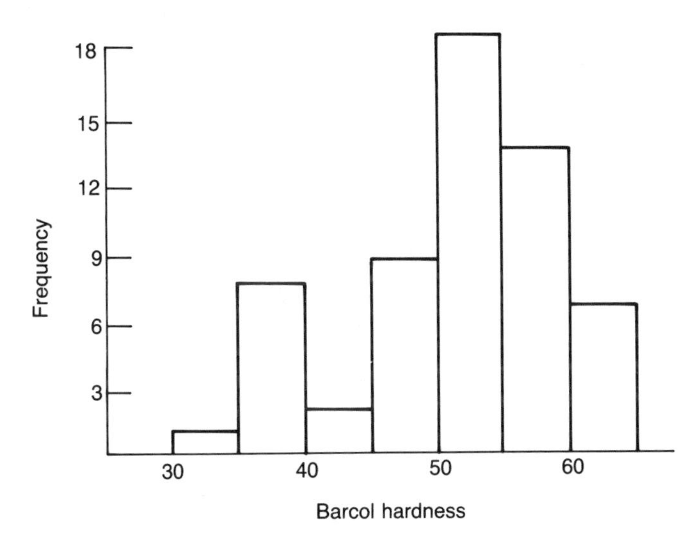

Figure 12.3 Barcol hardness of a moulded component (Mayer).

A further consideration is the shelf-life of constituents. This is typically three to six months for resins and 12 months for fibres and fabrics. If this is exceeded then at least one appropriate property needs to be measured to check that the material is still within specification. It is best to check this procedure for extending material life with the supplier.

For resins this property could be the gel time and for fibres the properties of the size or their interlaminar shear strength when incorporated into a laminate.

Sources

Mayer, R.M. (1985) Material selection and quality control of glass reinforced plastics, *Proc. 7th British Wind Energy Association Workshop, Oxford*, MEP, London.

ASTM D 2583–87 *Test Method for Indentation Hardness of Rigid Plastics by Means of a Barcol Impressor.*

EN 59: 1977 *Glass Reinforced Plastics – Measurement of Hardness by Means of a Barcol Impressor.*

■ 12.3 PROCESSING

■ **Quality system**

12.3 PROCESSING

The best strategy is to undertake an appropriate level of process control by checking each of the important steps during manufacture.

Moulded component

The moulded component shown in Figure 12.2 was manufactured by the RTM technique. The key steps were pre-forming the fabric reinforcement (section 11.3), inserting the fabric pack into the mould and injecting the resin.

A suitable method of process control is illustrated in Figure 12.4.

These steps control the process, most of which are self-explanatory. As thermosets undergo exothermic reactions on curing, the temperature rise on curing could be monitored, as an indication of the degree of cure and the time after which a component can be taken from the mould.

The quality of the component and the property level would be checked by use of both non-destructive and destructive test techniques (subsequent sections).

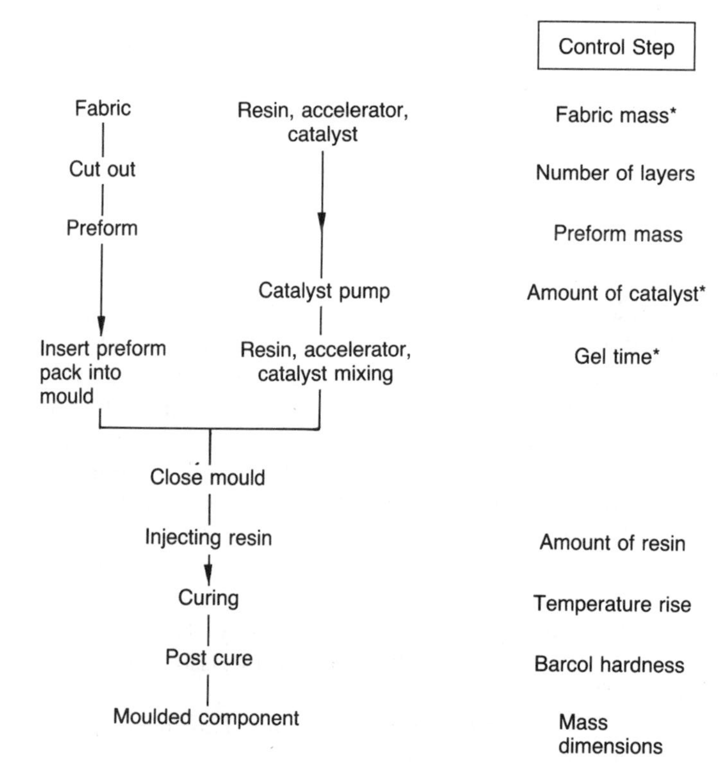

Figure 12.4 Process control of a RTM moulded component; steps marked * are to be undertaken at start of day or new batch of material.

Degree of cure

The degree of cure will vary with time and can be most easily monitored by the use of Barcol hardness (Figure 12.5).

In general a minimum value of Barcol hardness is needed to allow the component to be demoulded without distortion. Only experience will show what proportion of the manufacturers' recommended value for a fully cured resin this might be (generally around 40–60%).

For most applications, but particularly for those subjected to long-term stressing, the environment or the possibility of fire, a full cure is essential.

This may often require a post cure stage involving the application of heat. Again the Barcol hardness can be used to ensure that the hardness of the component complies at least with that value recommended by the manufacturer for the resin.

Other techniques include the use of differential thermal analysis.

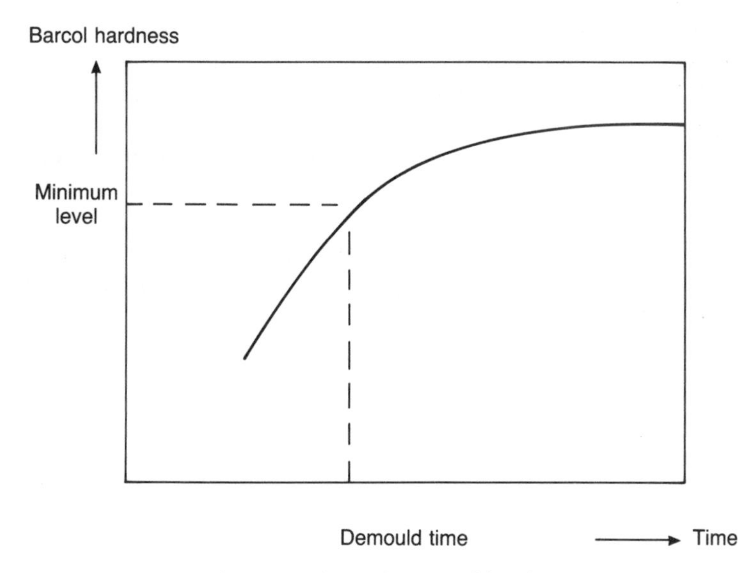

Figure 12.5 Monitoring the state of cure by Barcol hardness.

■ 12.3 PROCESSING

■ **Process control**

Hatch back door

The scheme developed for moulding the UNO hatch back door (section 11.4) is set out in Figure 12.6.

Observations

The production cycle time of three minutes is very short so the process control has to take this schedule into account.

The key steps are:

- a check of properties of incoming material;
- recording of the process data, which can be checked against the mean and coefficient of variation that is acceptable;
- optical and non-destructive testing of each component;
- destructive testing of a proportion of the assembled components;
- quality check of component after painting.

Design implications

Such a scheme is developed as the process is defined because quality control goes together with specific manufacturing steps.

The degree of automation is dependent upon the production rate; for a much longer production time, the data could for example be recorded by hand.

Such a scheme could be adopted for many types of components.

Source

Rossi, F. *et al.* (1987) Thermosetting composite injection and compression technology for a double wall rear door of a passenger car, *Proc. ANTEC Conference*, ANTEC.

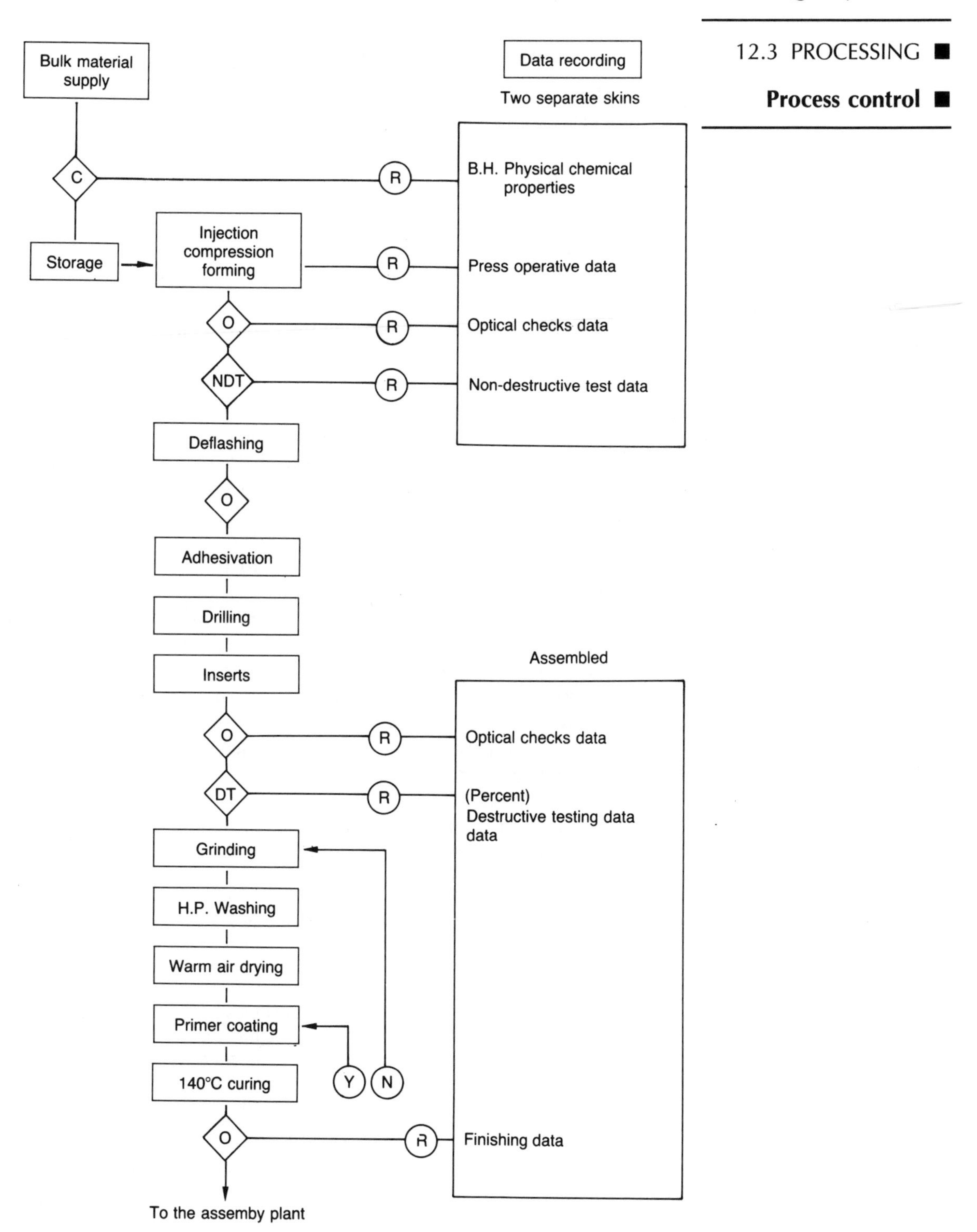

Figure 12.6 General layout of process and control for Fiat Uno Turbo door (Rossi *et al.*).

■ 12.4 NDE

■ **Techniques**

12.4 NON-DESTRUCTIVE EVALUATION

Such techniques are preferred as they can be carried out on actual components going into service and even during service to check their durability.

The major types are listed in Table 12.3.

We consider each of these briefly.

Visual check

This should always be done with reinforced plastics as a good indication of moulding quality. Many GRP components are translucent so one is able to assess the quality through the sample as well as that of the surface.

Dimensions

Shrinkage is dependent upon fibre volume as the fibres restrain the resin from shrinking. So this is a good indirect check of presence of reinforcement in the appropriate quantity at the right place.

Mechanical loading

This permits the stiffness of the component to be checked and compared with the design value.

It is important in many designs because fibre-reinforced plastics are generally stiffness limited compared with designs using metals.

Strength is harder to predict even with the data cited in this text. So mechanical loading to the design stress or slightly above will help to verify the calculated stress levels.

Strain gauges will permit local strains to be determined and thus the location of the higher stressed areas. These data can be compared more closely with design predictions.

Acoustic emission events can be detected on loading a component, but the type and number of events are difficult to relate to visible damage. One test method for chemical storage vessels compares the

Table 12.3 Principal non-destructive evaluation techniques

type	aspect, property
visual check	quality, surface finish
dimensions	design requirement
	quality control
Barcol hardness	degree of cure
	strength
mechanical loading to design stress	stiffness
	strength
acoustic emission on loading	component integrity
natural frequency, logarithmic damping	quality of fabrication
mechanical tapping	delamination of bonded parts
ultrasonics, X-ray scan	presence of cracks
infra-red imaging	possible areas containing damage

number of events on loading, unloading and reloading to the design load to determine whether any induced damage is stable or not. Sensors should always be located in highly stressed areas.

Other techniques

The fundamental frequency of vibration of the component gives an indirect indication of the quality of fabrication if this can be compared with the calculated value, for example, turbine blades.

Mechanical tapping is a simple and sensitive test for bonded structures to check the quality of the bond between parts.

Ultrasonics and X-ray scans are specialized techniques which can be used to detect the presence and location of damage under certain conditions.

Infra-red imaging has been used for structures undergoing repeated loading to detect damage areas and their rate of growth and so predict where failure might occur.

■ 12.4 NDE

■ **Fatigue testing**

NDE monitoring of fatigue testing

This is now becoming routine in coupons and components undergoing fatigue testing as it enables both the damage accumulation to be monitored and a failure criterion to be established. The example shown in Figure 12.7 is one in which the change in stiffness has been monitored.

Materials

- reinforcement – multiple ply UD/±45°;
- resin – iso-polyester (DSM resins).

Manufacture

Hand lay-up.

Testing

Constant amplitude at 20 Hz frequency, reverse loading tension/compression ($R = -1$), maximum stress 75 MPa, cycles to failure 23 million.

Observations

The change in stiffness has been monitored continuously over the life of the coupon. There appear to be three stages of degradation – an initial drop in stiffness, a long period of very slow change ('steady state') and a final stage leading to failure.

Design implications

Such a plot can be used to define an effective state of 'failure'. A sensible value from this plot might be a 10% drop in modulus if the design could sustain such a change in stiffness. If not then a smaller change would have to be adopted.

Such work forms part of a large international collaboration programme to determine damage accumulation and failure criteria. Most data are usually quoted as those at which a coupons fail to carry the applied load. A scale factor can then be applied to ensure that this does not happen. Direct measurements such as these would allow a failure criterion to be applied directly.

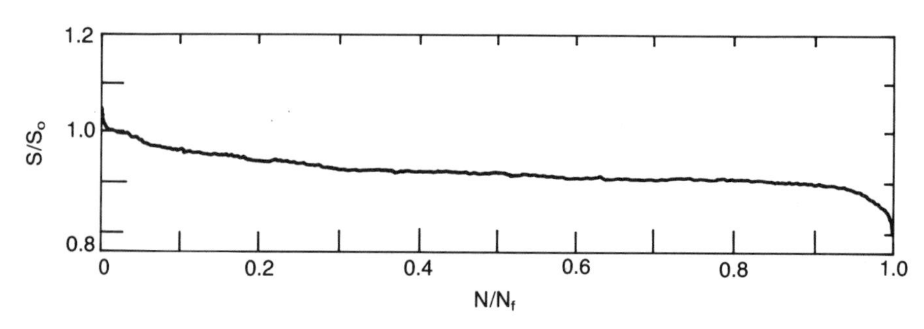

Figure 12.7 Degradation plot of a coupon undergoing fatigue testing normalized to the initial stiffness (S/S_0) and the number of cycles to failure (N/N_f) (Bach/ECN).

Source

Bach, P. (1988) High cycle fatigue investigation into wind turbine materials, *Proc. EWEC 88 Conference, Glasgow*, Stevens.

R.M. Mayer, coordinator of the CEC-funded Joule program on 'Fatigue and design of wing blades for wind turbines'.

■ 12.5 DESTRUCTIVE TESTING

■ **Techniques**

12.5 DESTRUCTIVE TESTING

These techniques give direct information about quality, property level and location of the reinforcement at the expense of destroying the component. As shown in Figure 12.6, a proportion of components have to be tested in this way especially when a component is in volume production.

The major types of test are listed in Table 12.4.

We consider each of these briefly.

Resin burn-off

The technique is only successful with glass, and it will show the location of the fibres. It is necessary to ensure that the fibre fraction and orientation are what the design requires them to be; of particular importance for short length fibres.

Mechanical loading to failure

This will allow the ultimate strength and strain and the failure mode(s) to be determined. Though strength is more difficult to predict than strain, the data in this text should allow an estimate to be made of the expected strength and strain.

If there is a large difference then there could be many reasons ranging from materials past shelf-life to incorrect processing or curing.

Strain gauges could be used to determine local strains in order to check design predictions. Failure modes are of particular importance in these materials as they can indicate why a component has failed.

Fire performance

There are little data available other than for flame spread (UL 94, limiting oxygen index). So the assistance of the resin suppliers must be sought at the start of the design. The tests described in Chapter 8 are only small-scale fire tests and would be suitable to monitor quality control of materials and components. Large-scale tests would be required to check

Table 12.4 Principal destructive testing techniques

type	aspect, property
resin burn-off	location of fibres fibre volume fraction (glass)
mechanical loading to failure	ultimate strength ultimate strain mode of failure
impact testing	energy absorption mode of failure at high strains
fire performance	rate of flame spread, heat build-up, smoke and toxicity
environmental influence	susceptibility to water ingress, chemical attack, wear or weathering
creep and fatigue	ability to withstand sustained load

the fire performance for use in a building or public space as a cladding material or partition.

Impact testing

Fibre-reinforced plastics are strain rate sensitive (Figure 6.6) so the ability to absorb energy will depend upon the loading rate. The advantage of drop weight impact tests (section 6.1) is that the indentor and its energy can be varied to simulate as far as possible the nature of the impact in service. Again a failure criterion can be selected depending upon whether penetration has to be prevented or surface damage is acceptable.

The data in Chapter 6 are indicative of what is available in the literature and it will be necessary to check the effect of impact experimentally.

Environmental influence

The concern is to simulate the conditions that will be encountered in service which could comprise more than one effect occurring at once. If there is any doubt about the validity of accelerated tests, one or more components should be exposed to the environmental influences, the testing being started ahead of any component being taken into service.

The data in Chapter 9 are only a small portion of what is known about chemical attack and so databases and material suppliers need to be consulted at an early stage in the design.

Creep and fatigue

There are sufficient data available to develop a suitable strategy, which subsequently needs to be verified. Coupons can be used to verify aspects of the design and performance in service, but ultimately full-scale tests would be required possibly on an on-going basis to monitor production quality.

■ 12.5 DESTRUCTIVE TESTING

■ **Scale of test**

Small- versus large-scale tests

The relative need will depend upon the application and the manner in which the materials are used. If the materials are highly stressed or exposed to long-term stressing, harsh environmental influences or fire, then full-scale testing must be undertaken.

As coupon testing can be used to provide quality control and property data these tests are the most frequent. Testing of sub-assembly and full-scale components or structures are the least frequent because of the complexity, cost and time to complete the tests. This gives rise to the 'pyramid of substantiation' as shown in Figure 12.8.

The relative costs of the various types of tests are also illustrated based on the development of a new fibre-reinforced plastic for an aircraft.

If only one of the process variables such as the base material is subsequently changed then the relative costs tend to be tilted towards coupon testing because spot checks can be carried out on the other properties.

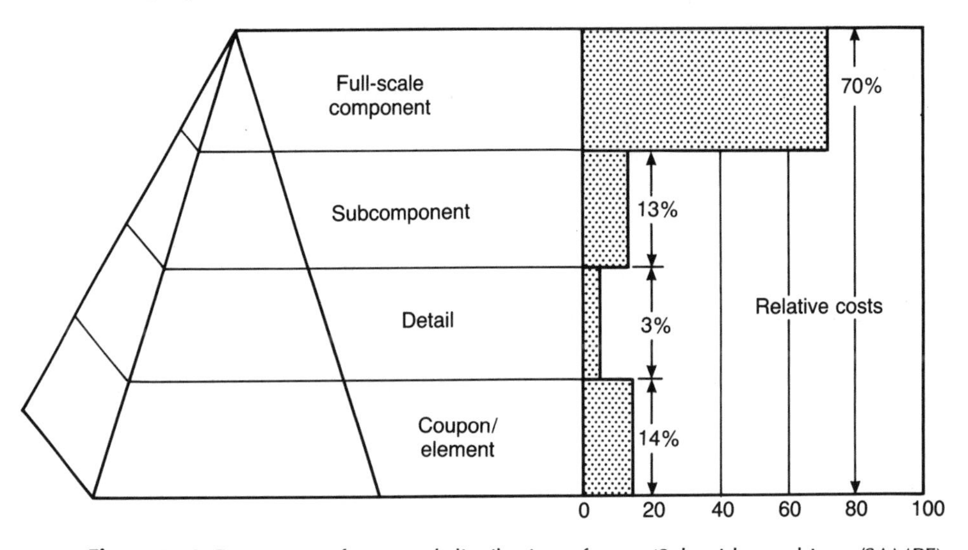

Figure 12.8 Frequency of tests and distribution of costs (Schneider and Lang/SAMPE).

Source

Schneider, K. and Lang, R.W. (1990) Second source qualification of carbon fiber prepregs for primary and secondary airbus structures, in *Proc. 11th International European Chapter Conference of SAMPE*, SAMPE.

Fatigue tests of blade materials

A comparison of the scale effects has been made between coupons and spars of wind turbine blades, which are subjected to a demanding fatigue load spectrum over a service life of 20 years (Figure 12.9).

Material

- reinforcement – UD and ±45° fabric lay-up;
- resin – L20/SL (Bakelite), or M9, M10 (Brochier).

Manufacture

Hand lay-up, test plates and prepreg spar beams.

Testing

Coupons through compression and tension/compression fatigue testing. Spars by fatigue testing in flapwise mode.

Observations

Data were collected on coupons laid up in the sequence specified for the spar as well as spars themselves. Testing was done initially with constant amplitude and then with a load sequence test characteristic of that experienced by wind turbine spars in service (Wispx).

A variety of NDE techniques was used including strain gauges, stiffness and, in addition for the spars, natural frequency and logarithmic damping. A change in natural frequency was detected as the testing proceeded (Figure 12.9).

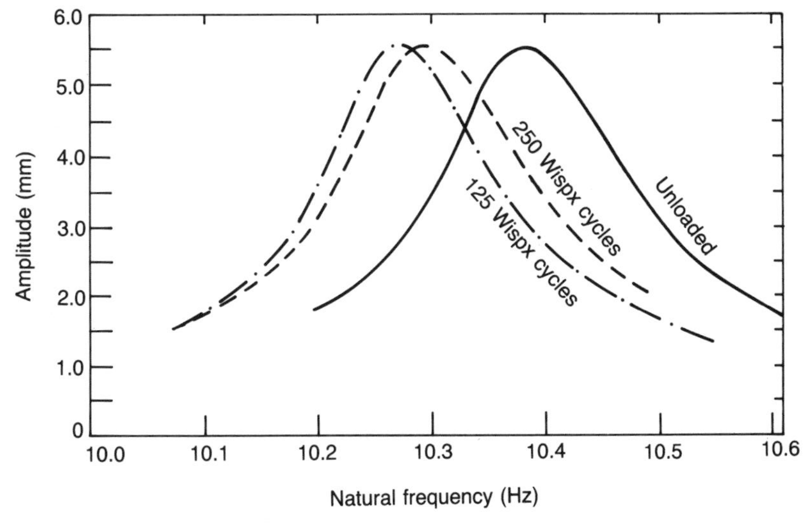

Figure 12.9 Frequency response for spar beam depending on life-time. Wispx is an amplitude spectrum which a blade will encounter in service (Kensche and Kalkul).

■ 12.5 DESTRUCTIVE TESTING

■ **Fatigue**

Natural frequency

Design implications

The failure modes were different in the spars (local buckling for example) than for the coupons. For the spars, visual examination was the most sensitive technique. The value of scaling up tests from coupons to components is clear.

Source

Kensche, C.W. and Kalkul, T. (1990) Fatigue testing of Gl-ep in wind turbine rotor blades, *Proc. EWEC 90 Conference*, Madrid.

12.6 STRATEGY FOR QUALITY CONTROL

A number of concepts has been discussed based on the ISO 9000 set of standards. A variety of tests has been briefly described and reference has been made to the data in this text where they are relevant.

The essential principle is that the design and process strategies have to be implemented side by side for load bearing components and structures. The principal design and process steps need to be capable of being checked; suitable test methods and process control have to be devised.

The flow charts for two different types of components illustrate how this can be achieved (Figures 12.4 and 12.6). Both the manufacturing process and the ultimate use of the component dictate what tests will be carried out and at what frequency. Some destructive testing will always be needed.

12.7 REFERENCE INFORMATION

General references

Quality Control in Composites Manufacture (1990) RAPRA Technology, Shawbury – papers given at a seminar outlining current and likely requirements of certification authorities and major companies.

Mayer, R.M. (ed.) (1993) *Design with Reinforced Plastics*, Design Council, London – both for general information and possible test methods.

Engineered Materials Handbook, Volume 1: Composites (1987) ASM International, Metals Park, Ohio – section 10, aspects of process control.

Phillips, L.N. (ed.) (1989) *Design with Advanced Composite Materials*, Design Council, London – Chapter 8, quality control and NDE of composite materials and components.

Pendleton, R.L. and Tuttle, M.E. (1989) *Manual on Experimental Methods for Mechanical Testing of Composites*, Elsevier, Barking – describes techniques primarily for NDE.

Summerscales, J. (ed.) (1987 to date) *Non Destructive Testing of Fibre Reinforced Plastics Composites*, Elsevier, Barking – series of review papers covering various topics.

Quality systems

Quality is considered in the ISO 9000 series of standards in the sense of fitness for purpose and safety in use. The specific parts are:

ISO 9000 – selection and use of the appropriate part of the series;
ISO 9001 – quality specifications for design, production, installation and servicing;
ISO 9002 – requirements for manufacturing goods or offering a service;

ISO 9003 – quality system for final inspection and test procedures;
ISO 9004 – overall quality management system and quality system elements.
BS 5750/ISO 9000: 1987 *A positive contribution to better business* (1990) BSI, Milton Keynes, 1990 – a brief guide to the use of the international standard for quality systems.

BSI, P O Box 375, Milton Keynes, MK 14 6LL, UK, Tel. (44) 0908 678 243, is one of a number of firms that can provide assistance with the implementation of this system.

INDEX

Page numbers appearing in **bold** refer to figures and page numbers appearing in *italic* refer to tables.